SpringerBriefs in Applied Sciences and Technology

PoliMI SpringerBriefs

More information about this series at http://www.springer.com/series/11159
http://www.polimi.it

Francesca Foglieni · Beatrice Villari
Stefano Maffei

Designing Better Services

A Strategic Approach from Design
to Evaluation

POLITECNICO
MILANO 1863

Francesca Foglieni
Department of Design
Politecnico di Milano
Milan
Italy

Stefano Maffei
Department of Design
Politecnico di Milano
Milan
Italy

Beatrice Villari
Department of Design
Politecnico di Milano
Milan
Italy

ISSN 2191-530X   ISSN 2191-5318 (electronic)
SpringerBriefs in Applied Sciences and Technology
ISSN 2282-2577   ISSN 2282-2585 (electronic)
PoliMI SpringerBriefs
ISBN 978-3-319-63177-6   ISBN 978-3-319-63179-0   (eBook)
https://doi.org/10.1007/978-3-319-63179-0

Library of Congress Control Number: 2017951055

Printed on acid-free paper

This Springer imprint is published by Springer Nature
The registered company is Springer International Publishing AG
The registered company address is: Gewerbestrasse 11, 6330 Cham, Switzerland

Preface

The argumentation presented in this book is the result of a research process initiated at the Design Department of Politecnico di Milano a few years ago, and matured thanks to a doctoral research developed by Francesca Foglieni, which constituted the starting point for the reasoning articulated in these pages. With this text, we share the urgent need to consider evaluation as a strategic activity that should become an integral part of service design, both in theory and practice. We believe it is necessary to support its role as a lever of growth and change in public and private organizations, and to design better services that answer to current social, economic and environmental challenges. This book relies on the idea that service design has the power to trigger meaningful changes, but for this to happen new directions need to be envisioned and explored for the discipline to evolve. Applied research projects, teaching activities and conversations with expert colleagues and professionals have also been fundamental to the full development of our vision. Our purpose is to open a debate on a topic that still deserves further investigations in the disciplinary field as well as in comparison with other disciplines. We hope that other researchers will share the hypotheses presented here and will prosecute the discussion with passion, dedication and relevant contributions.

Milan, Italy
June 2017

Francesca Foglieni
Beatrice Villari
Stefano Maffei

Contents

Chapter 1
Introduction

Abstract The introduction to the book traces the reasons and the intentions behind the choice of treating the topic of evaluation in conjunction with service design, as well as the urgency for such a debate to be open. An overview of the book structure and the reasoning articulated in the following chapters are provided and a reflection on upcoming topics concerning the future of the discipline is proposed.

Keywords Service design · Service innovation · Service evaluation · Open debate

Services are becoming the leading driver of the economy. Services are everywhere: they represent the main economic lever in the world countries (The World Bank 2017). Moreover, they are fundamental for all those issues regarding the welfare sphere and the social domains. As a consequence, service design is becoming increasingly important, in both the private and the public sphere: organizations, companies, and governments need to provide solutions that employ fewer resources, are more effective and are tailored to market and citizens' needs. Service companies and public administrations are including service design into their processes to create better services and to deliver better policies; agencies are empowering their competencies and skills toward a more service-oriented offering. Designing services is becoming an important issue, not to be considered only by a few practitioners. In this context, evaluating services is a way to design better and more efficient solutions, to redesign existing offerings, and to support better decisions, in order to create more social and economic value.

This book explores the connections between service evaluation, service innovation, and service design. Considering evaluation as a part of the service design process requires a renovated approach that includes new methods, tools, and competencies, and a new mindset to raise awareness of the contribution that evaluation can make to service innovation. Services are intangible and based on performances. They are heterogeneous due to the dependency of providers and users contexts. They perish after use while the consumption and the production are intrinsically connected (Zeithaml et al. 1985). These characteristics influence the evaluation process regarding metrics, outcomes, and strategies to be adopted.

© The Author(s) 2018
F. Foglieni et al., *Designing Better Services*, PoliMI SpringerBriefs,
https://doi.org/10.1007/978-3-319-63179-0_1

Furthermore, the scale and the diffusion of the service at hand also condition the way to determine its value.

In the first part of the book, the emphasis is on service value creation, considering both the provider and the user's point of view. Service evaluation, indeed, is mostly referred to the user side regarding, for example, customer satisfaction processes, often ignoring the provider's perspective or the interaction between the two. On the other hand, the evaluation of a service from the organization's point of view is mostly based on economic and quantitative parameters that rarely consider experiences and quality aspects such as interactions between people. This book proposes an overview of the evaluation theory and practice reflecting on the connections with service innovation and service design through an explorative path aimed at integrating service evaluation and service design into a single process. We suggest a theoretical perspective that initially outlines the key factors of the service economy by exploring the most important concepts of service innovation, and subsequently the central ones of the service design approach. After that, we introduce evaluation as a constitutive part of the design process, focusing on the importance of evaluating services to provide better design solutions.

To reinforce the concepts described and to transfer them into the service design practice, we analyze four cases—developed in the public and private sphere—to explain how evaluation is currently carried out within the service design process to assess existing services and service ideas. Lastly, a first reflection is done on how the integration of evaluation and service design can lead to a change of perspective for designing services as a professional practice in different areas.

Reflections on how to merge service evaluation and service design are just in their infancy. This book wants to contribute to turning on the debate by looking for connections that do not yet exist within the community of professionals and researchers, and formulate questions that will find a better-articulated response in the future. After more than two decades' evolution of the path of service design discipline, there is an urgency to better understand what is the impact that services have on people, organizations, and society. This work acts as the result of a collective reflection made through academic research activities, project planning and development in a constant dialogue with companies, agencies, institutions and people who have been working on service design and innovation for years.

There is a long road to be traveled. We cannot consider this book as a final result, but we hope it could be a step to strengthen further the idea of an evolving journey: we have just started, and we must continue to explore this territory. For this reason, we are aware of not being exhaustive, since important themes for evaluation and service design are just briefly mentioned, such as the relationship with co-design issues, the emergence of *design for policy* perspective and *impact evaluation*, to mention few.

Moreover, society is rapidly changing through new forms of economic and social transformations: new production and distribution models are emerging, and they will also affect the design, development and delivery of services. Examples are the robotization and the machine learning paradigms (Brynjolfsson and McAfee 2016), the digital transformation within the Industry 4.0 model and the so-called 4th

Industrial Revolution perspective (Schwab 2016). Reflecting on the effects that such and other changes induce on the service design sphere is not the purpose of this volume, but we hope that our reflection on evaluation in service design can be a stimulus for future discussions from both the theoretical and practical perspectives.

References

Brynjolfsson E, McAfee A (2016) The second machine age: work, progress, and prosperity in a time of brilliant technologies. Norton & Company, New York

Schwab K (2016) The fourth industrial revolution. Crown Business, New York

The World Bank (2017) Services, etc., value added (% of GDP). http://data.worldbank.org/indicator/NV.SRV.TETC.ZS. Accessed 4 June 2017

Zeithaml V, Parasuraman A, Berry L (1985) Problems and strategies in services marketing. J Mark 49:33–46

Chapter 2
From Service to Service Design

Abstract This chapter explores the origins of service design through an overview of key concepts and theories, starting from the definition of *service* and the characteristics of the service economy. Service marketing and service management literature are analyzed to describe the difference between products and services and to trace service peculiarities, which brought on the need of a dedicated design discipline and the formulation of the so-called *Service-Dominant Logic*, shifting the focus from services as goods to services as a perspective on value creation. The growing necessity and importance of designing services not only led to the birth of a new stream of study and practice in the field of design, through the development of specific methods and tools that support the creation of service solutions, but also allowed service design to become a crucial element for service innovation. Designing services that meet people's and organizations' needs, as well as the societal ones, is nowadays considered a strategic priority to support growth and prosperity. The final part of the chapter is therefore dedicated to outlining the role of service design in the contemporary socio-economic context as a driver for service, social and user-centered innovation.

Keywords Service · Service economy · Service sector · Service innovation · Service design · Service-Dominant Logic · Service value · Value creation · Design for service

2.1 Service: A Difficult Word

Since the first half of the 20th century scholars started to formulate a theory reflecting on changes in the economy and the division of labor (Clark 1940; Fisher 1939; Fourastié 1954). According to this theory, economies can be divided into three sectors of activity: extraction of raw materials (primary sector), manufacturing (secondary sector), and services (tertiary sector). We are probably aware of this distinction since primary school, but while it is quite clear to our minds what happens in the primary and secondary sectors, activities of the tertiary sector, also called the *service sector*, are harder to explain and interiorize.

© The Author(s) 2018
F. Foglieni et al., *Designing Better Services*, PoliMI SpringerBriefs,
https://doi.org/10.1007/978-3-319-63179-0_2

Despite services existing since the civil society and commercial activities were born, the conceptualization of *service sector* and its meaning, as we intend it nowadays, required a lot of time to be regarded as standard. As Allan Fisher asserted in 1952,

There was the implicit assumption that the classification into primary and secondary industries exhausted all the significant possibilities of employment. Everyone knew, indeed, that many people found employment outside the primary and secondary fields, in transport, in retail trade and other miscellaneous occupations. But these were regarded as rules of subordinate importance, and it was felt that there was no need for public policy, which actively concerned itself with both primary and secondary production, to pay much attention to them. (Fisher 1952: 822)

Service is a difficult word, and starting from the shift from the industrial to the service economy in the second half of the 20th century (Fuchs 1968) lots of definitions arose and are still arising. If we search for the word *service* in any dictionary, the first definition reads, "the action of helping or doing work for someone" or similar. Three keywords strongly emerge from this definition: *action*, *help* and *for someone*. From these, we can argue that a service is basically an activity done by someone in order to give support, satisfy the need of someone else. If we think of public services, for example, like healthcare or transportation, the concept is easily transferable: they are indeed activities provided by governments and public bodies to support individuals and communities in nursing infirm people and moving in the city.

The need of understanding and managing services brought by the shift toward a service economy, led to the emerging of specific disciplines (Fisk et al. 1993; Grönroos 1994), like service marketing and management, which struggled for a long time with the definition of service in order to enclose in a phrase the complexity and continuous evolution of the term.

In a first stance, many authors started describing services in comparison to products, identifying four properties that clearly distinguish the former from the latter. Based on a literature review conducted by Zeithaml et al. (1985), on a sample of 46 publications by 33 authors in the period 1963–1983, these characteristics, usually referred to as IHIP paradigm, consist of:

- Intangibility, because services are activities or performances rather than physical objects and there is no transfer of possession when they are sold;
- Heterogeneity, because every performance is unique since it depends on the behavior of the provider and the customer, and of other contextual aspects characterizing their interaction;
- Inseparability of consumption and production;
- Perishability or inability to inventory.

This paradigm has later been refused by service marketing scholars (Lovelock and Gummesson 2004), who considered these four characteristics not generalizable to all services, since

Replacement of human inputs by automation and rigorous application of quality improvement procedures have substantially reduced heterogeneity of output in numerous service industries. Outsourcing by companies and delegation by consumers to a specialist provider of tasks that they used to perform for themselves have greatly expanded the incident of separable services. And advances in information technologies and telecommunications, notably development of the Internet and digitalization of texts, graphics, video and audio, have made it possible to separate customers in both time and space from the production of numerous information-based services, thus destroying the twin constraints of both inseparability and perishability. [...] The underlying problems rooted in the extensive and still growing diversity of activities within the service sector. (Lovelock and Gummesson 2004: 32)

This statement well explains our difficulty in describing and treating services, since they are material in continuous transformation and evolution. Hence, with this book, we do not want to give a univocal vision on services and the service sector, seizing them in a definition that will probably change soon. We want instead to provide the knowledge and the instruments that allow people to understand them, as a raw material that can be shaped and governed. In the next section we explore the shift to the service economy in order to be acquainted with the growing role of services and their changing characteristics, and get prepared to design them.

2.2   The Service Economy

The transition from an agricultural to an industrial economy has been characterized as a "revolution". The shift from industrial to service employment, which has advanced furthest in the United States but is evident in all developed economies, has proceeded more quietly, but it too has implications for society, and for economic analysis, of "revolutionary" proportions. (Fuchs 1968: 2)

In his pioneering relation for the US National Bureau of Economic Research called 'The Service Economy', Victor Fuchs described for the first time the shift to the service economy, quantifying the reasons for this phenomenon. He asserted that since the end of World War II, the service sector had become the largest and the most dynamic element in the US economy, and most of the industrialized nations of the world started to follow the pattern set by the United States. The emergence of this country as the first service economy created a new round of priorities for economic research.

At the dawn of the 21st century, all highly industrialized countries have become *service economies* (Schettkat and Yocarini 2003). According to the United Nations Conference on Trade and Development (UNCTAD) the international trade in services was the primary driver of growth in 2014. International service exports accounted for 21% of total global exports in 2014, an increase of almost 5% compared with the previous year; while merchandise exports increased by only 0.3% in the year. Also referring to Gross Domestic Product (GDP) composition of developed countries,[1] the service sector contributes in the higher percentage. In 2015, the contribution of services[2] to the total GDP of the European Union was 71.2%, 62.4% at world level. To make some specific examples, for the US it counted 77.6%, for the UK 79.6%, for France 79%, for Germany 73.8%, for Italy 74.2%. And it is continuously growing.

The growing importance of the service economy has been justified in several ways since its arousal (Schettkat and Yocarini 2003). One reason could lie in changes in the demand-side of services, as a matter of needs satisfaction according to which services satisfy higher needs than goods. On the opposite, on the supply-side, manufacturing industries started to increasingly outsource their service activities to firms specialized in the provision of such services. Nowadays we can affirm that both components influenced the widespread importance of services in the evolution of the economy, but for long the debate focused on the differences between products and services and, as a consequence between manufacturing and service industries. Concepts like *servuction*, *servitization* and *Product-Service System* (*PSS*) have been coined to describe their relation, but always considering products to have a prominent role in services or as services to be intangible add-ons to products (De Brentani 1989; George and Marshall 1984; Langeard et al. 1986). It required a long time to services to gain their raison d'être, independently from a connection to goods.

The term *servuction* is a neologism created by Eiglier and Langeard in 1987, merging the word *service* and *production*. The concept represents the combination of material and human elements used to develop activities with the purpose of creating the service performance that an organization wants to propose to the market. According to Eiglier and Langeard's systemic model, the service results from the interaction of three basic elements: the customer, the product, and the contact personnel, who is coordinated in advance by the internal enterprise organization. Even if closely related to the presence of a product, the *servuction* concept introduces a major factor of future service design studies: the centrality of service interactions.

In the same period, Vandermerwe and Rada (1988) presented the term *servitization* in order to represent the shift toward the service economy of manufacturing companies. According to these authors, *servitization* consists of adding value to core corporate offerings through services. It is customer demand-driven and perceived by corporations as sharpening their competitive edges. A couple of decades later, in 2007, Andy Neely started collecting large-scale datasets on the actual *servitization* of manufacturing firms (Neely 2009) in order to demonstrate the consistency of the phenomenon in the contemporary panorama. Using publicly

---

[1]According to the CIA World Factbook, accessible online at https://www.cia.gov/library/publications/the-world-factbook/.

[2]Including government activities, communications, transportation, finance, and all other private economic activities that do not produce material goods.

available data drawn from the OSIRIS database,[3] he considered 10,000 companies (with over 100 employees) from around the world. The 2011 dataset revealed that the scale of *servitization* in manufacturing run at more than 30% (Neely et al. 2011), which means that 30% of manufacturing firms added services to their offer to the market.

Another concept that emerged from the shift to the service economy is that of *PSS*. This concept is a particular case of *servitization*, where the inclusion of a service offering into the corporate offering of manufacturing industries derives from the need of a lower environmental impact. The concept of *PSS* was firstly defined by Goedkoop et al. in 1999:

> A product service-system is a system of products, services, networks of "players" and supporting infrastructure that continuously strives to be competitive, satisfy customer needs and have a lower environmental impact than traditional business models.

It is interesting to notice that in this definition other key elements characterizing services emerge: the idea of *system* and the attention to *customer needs*. Despite their focus on the relation between products and services, these concepts clearly show the evolution toward a better understanding of services as a complex material to be managed by companies, which requires specific knowledge and specific tools.

To better understand services and how to design them, it must also be considered that the contemporary service economy is different from that of the 1950s when the shift started to be recognized and analyzed. Globalization, the digital revolution and the consequent transition to a knowledge society (Ghemawat 2011; Mulgan 2011) have transformed communication into a real-time communication all over the world, removing geographical boundaries. Further, they have influenced our needs, our ways of producing and consuming products and services, and the skills required to satisfy the quickly changing market (Agarwal and Selen 2011; Agarwal et al. 2015; Gallouj 2002). Moreover, the consequences of the 2008 financial crisis have placed extreme stress on governments and public bodies to maintain the current service offering, but also to introduce new welfare systems (OECD 2011; Mulgan 2015). This has further been exacerbated by complex societal problems that emerged before the recession (e.g. ageing populations, climate change, and spread of chronic illnesses), which still require both public servants and firms to do more with fewer resources available (Gillinson et al. 2010). To face these challenges, a permanent process of innovation needs to be established that continuously detects and adapts to emerging trends and requirements.

Before exploring the concept of service innovation and its connection to service design, it is important to provide some insights about those disciplines entirely dedicated to the management of service peculiarities. In the next section, we get in touch

---

[3]OSIRIS is a database of financial information, ratings, earnings estimates, stock data and news on global publicly listed companies, banks and insurance firms around the world. With coverage of over 125 countries, OSIRIS contains information on over 37,000 companies.

with these disciplines, which led to the formulation of the so-called *Service-Dominant Logic*, and to the foundation of the service design approach, as we know it today.

2.3   Service Marketing and Service Management: The Origins of Service Design

Service firms were among the first to observe the problems created by the old management and marketing structure (Grönroos 1994), but the interest in studying service-specific issues firstly emerged among marketing researchers. Scholars started developing new models, concepts, and tools based on services' peculiarities and their production and delivery processes during the 1970s. Among others, the doctoral dissertations by Judd (1965), Johnson (1969), and George (1972) offered a thorough description of the nature of services and specific problems in services marketing, while Wilson's (1972) and Rathmell's (1974) books on respectively professional services and the service sector were the first ones exploring marketing problems in service firms (Grönroos 1993). Three stages have been identified in the evolution of service marketing as a new academic field, from its embryonic beginnings in 1953 to its maturity in 1993 (Fisk et al. 1993). During the first stage (1953–1979), service marketing scholars struggled to assert the discipline's right to exist (Fisk et al. 1993; Swartz et al. 1992). This stage culminated in the debate on how and why services were different from goods (Shostack 1977). During the second stage (1980–1985) the *services versus goods* debate began to wane (Fisk et al. 1993). Scholars stopped asking if services are different from products and started reflecting on the implications of this difference, and the need for developing useful insights for marketing practices in the service sector (Lovelock 1983; Schneider 2000; Swartz et al. 1992). This stage also saw the first papers in new areas of investigation, such as service design (Shostack 1984), which flourished in the next period. During the last stage (1986–1993), service marketing became an established field within the marketing discipline (Fisk et al. 1993). Publications on topics like managing service quality given the heterogeneity of the service experience, designing intangible processes, managing supply and demand, and organizational issues merging marketing and operations functions (Fisk et al. 1993; Swartz et al. 1992) matured considerably. Service quality and customer satisfaction were two of the most studied themes also in the following period.

Nevertheless, a radical reformulation of service marketing only happened in 2004, when Vargo and Lusch published the paper 'Evolving to a New Dominant Logic for Marketing' for the *Journal of Marketing*. The main conclusions of this article, as better explained in the next section, was that service must be considered more a perspective on value creation than an activity or a category of market offerings (Edvardsson et al. 2005). The work of Vargo and Lusch has framed the results of 30 years of service marketing research into one organized structure (Grönroos and Ravald 2011), which contributed in defining implications of adopting a service perspective for management and paving the way to a widespread recognition of the service design approach.

The service management perspective has emerged from several disciplines: marketing, operations management, organizational theory and human resources management, management, and service quality management, and established itself as a recognized discipline as well (Grönroos 1993). It is a perspective that supports firms involved in service competition and that have to understand and manage service elements in their customer relationships to achieve a sustainable competitive advantage. A definition of service management by Albrecht (1988: 20) well represents the shift from scientific management principles:

> Service management is a total organizational approach that makes quality of service, as perceived by the customer, the number one driving force for the operations of the business.

Before the arousal of the service economy, mass production and economies of scale were considered fundamental parts of management. Pioneering scholars in service management thought instead that the nature of the customer relationships and operations, and the production and delivery processes were different for services (Grönroos 1982; Normann 1982), and that applying

> a traditional management focus on cost reduction efforts and scale economies may become a management trap for service firms and lead to a vicious circle where the quality of the service is damaged, internal workforce environment deteriorates, customer relationships suffer, and eventually profitability problems occur. (Grönroos 1993: 8)

An entirely new approach to the management of various service organization aspects, and how relationships within the organization and between the organization and the customer or other stakeholders should be viewed and developed, was undertaken, in a way that cannot be separated from marketing findings (Grönroos and Gummesson 1985). The introduction of service marketing and management perspectives contributed to the formulation of concepts such as customer participation in the production and delivery process, co-creation of value, and the idea of a holistic approach, which later brought about the birth of service design.

2.3.1   From Services as Goods to Services as a Perspective on Value Creation

As stated before, the way in which services are viewed, considered and treated has changed over time. The traditional economic worldview before the arousal of the service economy was based on the so-called *Goods-Dominant Logic* (Vargo and Lusch 2004, 2008, 2014). This paradigm remained unquestioned for a long time, thus, when services began receiving attention from academics and practitioners, they were treated as intangible add-ons to goods or a particular type of immaterial product (Vargo and Lusch 2014). These goods-centered views of services were also evident in the four characteristics used to distinguish services from goods (see IHIP paradigm in Sect. 2.1) (Parasuraman et al. 1985). These features were generally considered to be disadvantages of services, thus requiring to be adjusted according to principles of marketing and management of goods provision. As previously discussed, problems caused by this view made scholars question the development of a distinct subdiscipline dedicated to services within marketing and management boundaries. This also brought on discussion of the nature of service relationships (Grönroos 2000), and a rethinking of the concept of service quality, how the customer perceives it and how it can be measured. In the words of Vargo and Lusch (2014: 44),

> Service quality moved the focus of the firm from engineering specifications of goods production to the perceived evaluations of the customers. Relationship marketing shifted the focus of exchange from discrete transaction to ongoing interactivity.

Together with the adoption of service subdisciplines by marketing and management, other signs of transition toward a full awareness of service peculiarities became evident, such as the move from a commoditized mass production to *mass-customization* (Pine and Gilmore 1999). Or the shifting of the focus of value creation from the firm to the customer, thus affirming for him/her a new and active role in service provision (Prahalad and Ramaswamy 2004). Or again the transition towards an *experience economy* that goes even beyond the service one (Pine and Gilmore 1999).

In 2004, all these insights were translated by Vargo and Lusch into a new marketing logic that they called *Service-Dominant Logic*. This new paradigm inverted the role of services in business and economic exchange: service, defined as the application of resources for the benefit of another actor (Vargo and Lusch 2004), is now considered the basis for economic exchange, where goods become a medium of service provision. Moreover, service value is always co-created by a service provider and a beneficiary, and no more by the sole provider during the production and distribution processes. The customer integrates his/her knowledge and capabilities with those of the firm. This understanding of service changed the conceptual position of the user from being a passive consumer, exclusively involved in the moment of purchase, to an active co-creator of value. *Service-Dominant Logic* also enhanced the understanding of value created in use and context (Chandler and Vargo 2011; Vargo and Lusch 2008), reinforcing the emphasis on service as a perspective on value creation rather than a replacement of products.

Recently, the interest in the understanding of user involvement in the service development process has emerged (Ostrom et al. 2010), and a critical stream of *Service-Dominant Logic* has developed bringing forward the position that it is the organization that takes part in the customers' co-creation of value rather than the other way around (Grönroos 2008; Heinonen et al. 2010). Thanks to this overview on the development of a service culture, we can now explore how it contributed to the formulation of the service design approach, within the traditional product-oriented design boundaries.

2.3.2   Core Concepts in Service-Dominant Logic and Their Relevance for Service Design

Despite their shared ambitions in improving customers' lives with better products and services, and being both customer-centric, traditionally, marketing and design have had a troubled relationship (Bruce 2011; Holm and Johansson 2005). The first often considered the latter only a tool or method in the marketer's toolkit (Bruce and Bessant 2002). Nevertheless, as well as for marketing and management, in the last decades, designers and design researchers have approached services as new possible objects of design (Blomkvist et al. 2010; Meroni and Sangiorgi 2011; Pacenti and Sangiorgi 2010; Sangiorgi 2009), enhancing the growth of a new strand of design competencies.

To face the development of new services, service marketing usually focuses on the service delivery and consumption moment. On the contrary, designers focus on user involvement and a thorough understanding of the context in which the service takes place. For this reason, service design is usually described as a holistic approach that recognizes relations and interactions characterizing a system (Mager 2009; Manzini 2009; Sangiorgi 2009; Stickdorn and Schneider 2010). With the development of the *Service-Dominant Logic* perspective and the description of service as a process in which users actively participate in the creation of value through interactions with the service provider, the role of design has then become to understand how actors relate and act within this system for value creation (Wetter-Edman 2014). Service design practices, previous to the formulation of the *Service-Dominant Logic*, were often put in relation to other design disciplines such as interaction and industrial design (Wetter-Edman et al. 2014), comparing for example service interactions and interaction design, in order to justify the adoption of tools and concepts from this field (Holmlid 2007). This vision emerged from understanding service as a category of market offerings rather than a perspective on value creation as discussed before. The implications of *Service-Dominant Logic* on service design have been widely discussed (Hatami 2013; Kimbell 2011; Segelström 2010; Wetter-Edman 2011). The overlap of key concepts in *design thinking* and *Service-Dominant Logic* have been explored and found complementary rather than overlapping, such as the understanding of value as *value-in-use* and *value-in-context*, experience as individually determined, and networks and actors as relevant players in the value creation process (Vargo and Lusch 2016). A discrepancy was instead found in the role of people in the value creation process, since for design they are seen as active users, while for the *Service-Dominant Logic* they are passive customers, at least in the first version of 2004. However, as briefly stated in the previous section, more recent interpretations of the *Service-Dominant Logic* (defined as *Service Logic*) by scholars like Grönroos and Voima (2013) make the two approaches align also on this point. Grönroos and Voima (2013: 138) argue that, as opposed to traditional descriptions of value creation and co-creation that place the firm in control of value creation (only inviting the customer to join the process as co-creator), the customer both creates and evaluates value over time, in an *experiential process of usage*. They assert that

> In the same way that the firm controls the production process and can invite the customer to join it as a co-producer of resources, the customer controls the experiential value creation process and may invite the service provider to join this process as a co-creator of value. (Grönroos and Voima 2013: 138)

Based on this argument they further suggest that the service provider should consider how to be involved in the customers' lives instead of getting the customers involved with their business.

Supporting the link between the two approaches, frameworks integrating design and the *Service(-Dominant) Logic* perspective have been recently developed, such as the *design for service* approach by Kimbell (2009), and Meroni and Sangiorgi (2011). Differences between *service design* and *design for service* will be explored later. An interesting contribution from the *design for service* approach, useful to understand the contribution of the *Service-Dominant Logic* to service design, is the focus on the *value-in-use* concept, especially when related to *value-in-context*, where context is defined as "a set of unique actors with unique reciprocal links among them" (Chandler and Vargo 2011: 40). In fact, Chandler and Vargo (2011) argue that for designing and managing services, it is necessary to deepen our understanding of contexts and its heterogeneous and distinctive nature.

Thanks to this exploration of the origins of service design and its connections with other approaches within the service culture, we can now deal with the important concept of service innovation, to then understand the importance of the discipline in the evolving service society.

2.4   The Emerging Interest in Service Innovation

Like the service concept, the innovation concept has remained for a long time mainly related to manufacturing-based paradigms (Gallouj and Weinstein 1997), whereas other forms of innovation embedding more intangible elements are still underexplored.

In recent reports, the European Commission underlines the fragmentation of the European market where, although the service sector represents 70% of the economy, knowledge-intensive services are still underdeveloped (European Commission 2010; OECD 2005). However, also in this case, a paradigm shift is peeking out, demonstrated by the strategic document 'Europe 2020 Flagship Initiative Innovation Union', where we can read that Europe is investing in service innovation by

> pursuing a broad concept of innovation, both research-driven innovation and innovation in business models, design, branding and services that add value for users and where Europe has unique talents. (European Commission 2010: 7)

Many scholars agree that innovation in service firms differs from innovation in manufacturing firms (Johne and Storey 1998; OECD 2000) and that it is often non-technological. The study of technical change in the service sector was largely neglected since services were seen as low-technology products (Cainelli et al. 2004; Lopes and Godinho 2005). In the innovation ecosystem, service companies have since been considered as facilitators, and services described as passive reactors to innovation taking place in the manufacturing sector (OECD 2000). The debate on the distinction between service and manufacturing innovation is still open, and some essential characteristics have emerged (Howells and Tether 2004; Salter and Tether 2015; Tether 2003, 2013).

Service innovation does not require as much R&D, and being immaterial it is simpler to imitate (OECD and Eurostat 2005). For this reason, service firms do not invest in patents or licenses. Innovation in services involves transformation concerning how the service is designed and developed, to how it is delivered and managed (Miles 2010; Trott 2012). It is a mix of product and process innovation and entails new ways in which customers perceive and use the service. Technology is often the servant rather than the core of new service development. Services have a higher degree of customization and changes include *soft dimensions* (social innovations, organizational innovations, methodological innovations, marketing innovations, innovations involving intangible products or services) beyond traditional *hard* technological-driven innovation practices (Djellal and Gallouj 2010). According to Van Ark et al. (2003: 16) service innovation can be defined as

> A new or considerably changed service concept, client interaction channel, service delivery system or technological concept that individually, but most likely in combination, leads to one or more (re)new(ed) service functions that are new to the firm and do change the service/good offered on the market and do require structurally new technological, human or organizational capabilities of the service organization.

Thus, it can be said that innovation in services involves different aspects (Fig. 2.1) concerning the development or improvement of service concepts, interfaces/touchpoints, delivery systems, or adopting a new technology (den Hertog 2000). This means providing a new or a renewed offering to suppliers or customers, producing benefits for the provider organization, defining new business models, and designing new ways of interaction or valuable customer experiences.

But what about service innovation today? In the current networked world, service innovation is highly interactive and systemic in nature, since both public and private service organizations are embedded in wide value networks that include suppliers, intermediaries, customers and partners, and that combine their capabilities in co-creation processes (Agarwal et al. 2015). Entities in these networks connect through human or technical channels, highlighting the importance of both human-centricity and technology in contemporary services. This interactivity offers organizations more opportunities and abilities to deliver valuable services resulting in service innovation (Agarwal and Selen 2011).

Although contemporary economies are undeniably service economies, and services have finally gained the right acknowledgement over products, the debate

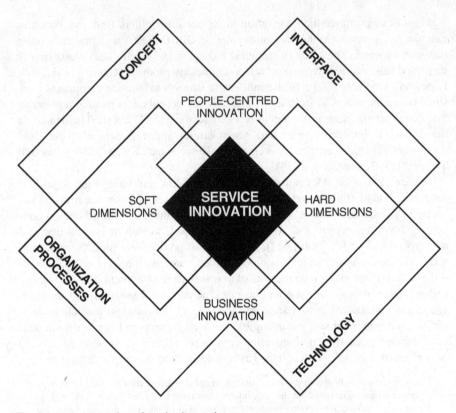

Fig. 2.1 A representation of service innovation

around service innovation is still alive. In this context, understanding the role of service design as a driver of innovation becomes critical.

2.5 Why Do Services Need Designing?

Even though service design can be considered a new field of expertise, it founds its roots in a vast spectrum of disciplines and its history can be referred at least to the past 30 years. As shown in Fig. 2.2, scholars are unanimous in considering service design as a multidisciplinary practice (Meroni and Sangiorgi 2011; Miettinen and Valtonen 2012; Moritz 2005; Stickdorn 2010; Stigliani and Fayard 2010) that integrates different approaches and tools derived, among others, from psychology, marketing and management, IT and interaction, user-centered and graphic design. This characteristic makes it possible for service designers to face the service development process, as well as the design of both intangible interactions and physical elements (Moritz 2005).

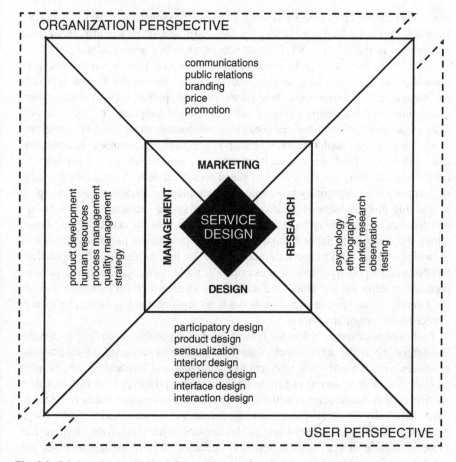

Fig. 2.2 Service design multidisciplinary nature (elaboration from Moritz 2005)

At the beginning of the 1980s, Lynn Shostack started talking about the design of immaterial components and introduced the *service blueprint* as a design tool (Shostack 1982, 1984). But the official recognition of service design as a distinct disciplinary field of investigation only occurred in 1991 when Michael Erlhoff and Birgit Mager introduced it at the Köln International School of Design. From that moment forward, service design has been deeply explored both in the academic and professional field. Books and papers on this topic began to be published, discussing definitions, competencies of service designers and service design tools (Mager and Gais 2009; Polaine et al. 2013; Stickdorn and Schneider 2010). Also, the number of service design agencies has increased since then, reinforcing the link between service designers' specific skills and business. Moreover in 2004, the Service Design Network—an international network of organizations and institutions working in the domain of service design—was created, and many service design university programs have been structured worldwide. As a matter of fact, many

European countries provide service design training programs nowadays (i.e. in UK, Denmark, Finland, Germany, Italy, Switzerland), and there are also interesting experiences in the US (i.e. NY Parsons School, SCAD, Carnegy Mellon).

The current panorama looks promising for service design: research and experimental projects on services, service innovation, and the service design approach are conducted in universities, businesses and the public sector. Furthermore, international service design agencies are increasingly supported by local service design agencies in facing the complexity of the contemporary context where services (for example health services, transports, banking, insurances, communications, education, food services, cultural services, tourism, etc.) have become the crucial ingredients of our daily life. As stated above, services represent a significant segment of the contemporary economic and technological landscape, and they play a vital role in knowledge-based economies. New service companies are growing, and services are becoming more and more important also for manufacturing firms, where the trend is to involve final users in the production processes, thanks to technology-based support services that enhance the user experience. Best practices like Nike and Apple represent clear examples of the service capacity to add value to products, reinforcing the brand and making the client a faithful customer through the offering of immersive experiences made by multichannel systems of product, services and communications.

From this widespread acknowledgment, we can argue that a growing awareness is establishing in the service sector: services need to be designed. Organizational processes, creative activities, interactions, marketing and business strategies, user analysis and user research, use of technologies, and prototypes are all parts of a service system that requires multidisciplinary contributions and needs to be managed holistically by an all-encompassing professional figure (Moritz 2005). Large companies, SMEs, associations and public authorities are recognizing it more and more in service design practitioners and agencies. Service design involves the capability of connecting the needs of customers with those of the organization, improving the quality of experiences, and supporting the organization in creating value, reducing the delivery gap (Allen et al. 2005), and differentiating from competitors (Moritz 2005). This capability does not only concern the development of new services, but also service redesign (Berry and Lampo 2000). A recent study by Sangiorgi et al. (2015: 62–63) confers in three particular contributions of service design to service innovation:

1. Service design as a skilled contribution to address a specific need;
2. Service design as a people-centered, creative and systematic process;
3. Service design as a collaborative and people-centered mindset and approach.

People-centeredness clearly emerges as an essential characteristic of service design (Holmlid 2009; Meroni and Sangiorgi 2011; Stickdorn 2010), since services are co-produced between people and providers, and they result from complex interactions inside and outside the service organization (Polaine et al. 2013). To sum up, thanks to the adoption of a service design approach, service organizations can achieve service innovation by creating or redesigning services that are closer to user needs, providing tailored experiences that are more easy-to-be-accessed by the customer, and giving coherence and consistency to all the elements of a service. This means to increase its desirability and usability, but also efficiency and effectiveness of organizational performances (Polaine et al. 2013).

2.5.1 Service Design and Design for Service

Over the past few years, together with the service design approach, also the concept of *design for service* (Kimbell 2011; Kimbell and Seidel 2008; Meroni and Sangiorgi 2011; Sangiorgi 2012; Wetter-Edman et al. 2014) has emerged. We find it relevant to spend few words in shedding some light on this subtle distinction. *Design for service* is proposed as a context-related approach to service innovation (Kimbell and Seidel 2008; Meroni and Sangiorgi 2011; Wetter-Edman et al. 2014), based on the *Service-Dominant Logic* analytical framework. According to Meroni and Sangiorgi (2011: 10),

> While acknowledging service design as the disciplinary term, we will focus more on articulating what design is doing and can do for services and how this connects to existing fields of knowledge and practice.

In their vision, which is also supported by Manzini (2011), the use of the preposition *for* implies the idea of transformation, the idea of designing as a transformative process. They do not consider a service as an object, but as a platform for action that can enable different actors in collaborating and be actively engaged in complex systems that involve a multiplicity of interactions. *Design for service* accepts

> the fundamental inability of design to completely plan and regulate services, while instead considering its capacity to potentially create the right conditions for certain forms of interactions and relationships to happen. (Meroni and Sangiorgi 2011: 10)

Kimbell (2011) explores different ways of approaching service design, including *design for service*. She compares ways of thinking about design (design as enquiry and design as problem-solving) with ways of thinking about service (when distinctions between goods and services are maintained and when service is considered the core unit of economic exchange). From this comparison, a framework made of four quadrants emerges. In her vision, *design for services* resides at the intersection between design as enquiry and service as the basic unit of economic exchange. By referring to *design for service*, rather than service design, she argues "the purpose of the designers' enquiry is to create and develop proposals for new kinds of value relation within a socio-material world" (Kimbell 2011: 49). In practice, this means that designers do not only work to produce a deliverable for the firm, but they rather work for creating materials and activities aimed at involving actors from the organization (managers, customers, personnel, etc.) in the enquiry. In this sense, *design for service* can be viewed as a particular kind of service design.

This contribution has been further expanded by Wetter-Edman et al. (2014), through the analysis of the contribution of the *Service-Dominant Logic* to *design for service* and vice versa. In agreement with Kimbell, but putting more emphasis on users and value co-creation, she considers *design for service* as focused on "observing and understanding users, at the time and places where value is co-created" (Wetter-Edman et al. 2014: 5), and designers working for service as facilitators of co-design and co-creation processes. Three propositions (plus one, which is not relevant for our argumentation) are described to frame the *design for service* approach:

1. *Design for service* explores existing service systems in order to design new services, through the understanding of different actors' perspectives and the value co-creation activities in which they are involved;
2. *Design for service* provides the mindset, together with tools and competencies, to understand how experiences are formed in contexts from different actors' perspectives;
3. *Design for service* extends the meaning of value co-creation through the adoption of co-design approaches that enable the creation of new service systems.

According to the exploration of *design for service* literature, it can thus be argued that, while service design is usually referred to as the overall discipline and practice of designing service, *design for service* represents the investigative (or research) and preparatory side of service design, focused on the understanding of the context and the design of activities aimed at people involvement. In particular, the term *design for service* puts emphasis on the complex and relational side of services, considering them as entities that are impossible to predetermine (Meroni and Sangiorgi 2011; Sangiorgi and Prendiville 2017). In the most recent publication on the topic, Sangiorgi and Prendiville (2017) adopt the term *designing for service* to indicate the importance of a context-aware approach to service design.

> Being 'designing' an ongoing activity to which designers can engage with and affect during their interventions, the focus necessarily shifts to the context of where these changes can and are happening, which is no longer exclusively just the user's space, but also the organisations and their value networks. (Sangiorgi and Prendiville 2017: 252)

This vision highlights the necessity for service design to create (and measure) service value both from the user and the provider point of view in a specific context, which is one of the inferences that will also guide the contents of this book.

2.6 The Strategic Use of Service Design in (Public and Private) Organizations

In the last decade, service design has started being considered as a key driver for service innovation, social innovation and user-centered innovation (European Commission 2010). Reinforcing the idea that the service design approach is a fundamental element for introducing change in business, organizations and the public sector, the necessity to tailor solutions that better fit users' needs is becoming a central issue. Businesses need to exploit their products/services offering to provide more involving experiences to their clients. Organizations need to create more and more tailored solutions to be able to address different users and communities. Public bodies need to respond to the increasing demand of high-quality solutions, to transform decision-making processes into a participative journey, and to deliver effective activities while optimizing the use of resources.

Service design can support all these entities providing tools and techniques that allow them to creatively introduce new concepts and solutions, focusing in particular on user research, and visualization methods and prototypes that enable to share ideas among different actors. This means making visible intangible elements such as relationships, experiences, as well as soft qualities of the environment. Moreover, the adoption of the service design approach can help people within organizations think like designers, which means supporting their capacity to use creativity, transform tacit knowledge into explicit ideas, and brace oneself in listening to users and collaborate with them. From the user perspective, interaction with services is often characterized by frustration: they feel ignored or misunderstood, they spend a lot of time to receive answers, and they face situations that affect the perception of performances as inefficient. For these reasons, both in public and in private organizations, trust mechanisms and customer satisfaction are key elements to work on. People and society change faster than organizations: this implies that service providers need to adopt quicker ways for understanding people's perspective and increasing customer satisfaction. What service design can do is helping them shape their offering according to user needs and expectations.

Some challenges can be recognized as drivers for delivering better services: for example, ICT and social media are introducing new ways of collaboration and co-creation of services. Taking the public sector as a reference, an excellent opportunity to renew the service offering is available for managers and citizens activating processes that are more citizen-centric, collaborative and networked. The idea of openness is crucial in the current service design era, where open data and digitization are important issues to be always taken into consideration. Open and participative legislative processes, community-based initiatives and user-oriented performances constitute the core of efficient and effective public and private organizations. Thanks to co-design practices, service design can contribute to developing new solutions, selecting the most promising ones and prototyping them involving citizens, businesses, intermediaries and institutions (Bason 2014).

We conclude this first chapter with a reflection by Thackara (2005) that well represents the role of service design in the contemporary socio-economic context. He states that the secret for innovating is the re-combination of different types of expertise productively, namely adopting a new kind of design able to increase the flow of information within and between people, organizations and communities to stimulate continuous innovation among groups of individuals within continuously changing contexts. Accordingly, service design focuses on the development of solutions that are coherent to users' satisfaction and well-being, on supporting organizations and communities to build a shared vision of transformation, on the adoption of tools to co-create and prototype services, proposing new approaches that foster changes in organizations and society.

References

Agarwal R, Selen W (2011) Multi-dimensional nature of service innovation: operationalisation of the elevated service offerings construct in collaborative service organisations. Int J Oper Prod Man 31(11):1164–1192

Agarwal R, Selen W, Roos G, Green R (eds) (2015) The handbook of service innovation. Springer, London

Albrecht K (1988) At America's service. Dow Jones-Irwin, Homewood

Allen J, Reichheld FF, Hamilton B, Markey M (2005) Closing the delivery gap. http://bain.com/bainweb/pdfs/cms/hotTopics/closingdeliverygap.pdf. Accessed 3 Mar 2016

Bason C (2014) Design for policy. Gower Publishing, Aldershot

Berry LL, Lampo SK (2000) Teaching an old service new tricks: the promise of service redesign. J Serv Res 2(3):265–275

Blomkvist J, Holmlid S, Segelström F (2010) Service design research: yesterday, today and tomorrow. In: Stickdorn M, Schneider J (eds) This is service design thinking. BIS Publishers, Amsterdam, pp 308–315

Bruce M (2011) Connecting marketing and design. In: Cooper R, Junginger S, Lockwood T (eds) The handbook of design management. Berg Publishers, Oxford, pp 331–346

Bruce M, Bessant J (2002) Design in business. Strategic innovation through design. Prentice Hall in cooperation with Design council and Financial Times, London

Cainelli G, Evangelista R, Savona M (2004) The impact of innovation on economic performance in services. Serv Ind J 24(1):116–130

Chandler JD, Vargo SL (2011) Contextualization and value-in-context: how context frames exchange. Mark Theory 11(1):35–49

Clark C (1940) The conditions of economic progress. Macmillan, London

de Brentani U (1989) Success and failure in new industrial services. J Prod Innov Manage 6(4):239

den Hertog P (2000) Knowledge-intensive business services as co-producers of innovation. Int J Innov Manage 4:491–528

Djellal F, Gallouj F (2010) Services, innovation and performance: general presentation. J Innov Econ 5(1):5–15

Edvardsson B, Gustafsson A, Roos I (2005) Service portraits in service research: a critical review. Int J Serv Ind Manage 16(1):107–121

Eiglier P, Langeard E (1987) Servuction. Le Marketing des services. Wiley, Paris

European Commission (2010) Europe 2020 flagship initiative innovation union. https://ec.europa.eu/research/innovation-union/pdf/innovation-union-communication_en.pdf. Accessed 18 Dec 2016

Fisher AGB (1939) Production, primary, secondary and tertiary. Econ Rec 15(1):24–38

Fisher AGB (1952) A note on tertiary production. Econ J 62(248):820–834

Fisk RP, Brown SW, Bitner MJ (1993) Tracking the evolution of the services marketing literature. J Retail 69:61–103

Fourastié J (1954) Die große Hoffnung des 20. Jahrhunderts, Köln-Deutz

Fuchs V (1968) The service economy. Columbia University Press for National Bureau of Economic Research, New York

Gallouj F (2002) Innovation in the service economy: the new wealth of nations. Edward Elgar Publishing, Cheltenham

Gallouj F, Weinstein O (1997) Innovation in services. Res Policy 26:537–556

George WR (1972) Marketing in the service industries. Unpublished dissertation, University of Georgia

George WR, Marshall CE (1984) Developing new services. American Marketing Association, Chicago

Ghemawat P (2011) World 3.0: global prosperity and how to achieve it. Harvard Business Review Press, Boston

Gillinson S, Horne M, Baeck P (2010) Radical efficiency—different, better, lower cost public services. http://innovationunit.org/sites/default/files/radical-efficiency180610.pdf. Accessed 23 Jan 2016

Goedkoop MJ, van Halen CJG, te Riele HRM, Rommens PJM (1999) Product service systems: ecological and economic basics. http://teclim.ufba.br/jsf/indicadores/holan%20Product%20Service%20Systems%20main%20report.pdf. Accessed 21 Jan 2016

Grönroos C (1982) Strategic management and marketing in the service sector. Swedish School of Economics and Business Administration, Helsingfors

Grönroos C (1993) Toward a third phase in service quality research: challenges and future directions. Adv Serv Mark Man 2(1):49–64

Grönroos C (1994) From scientific management to service management. A management perspective for the age of service competition. Int J Serv Indust Manage 5(1):5–20

Grönroos C (2000) Service management and marketing: a customer relationship management approach, 2nd edn. Wiley, New York

Grönroos C (2008) Service logic revisited: who creates value? And who co-creates? Eur Bus Rev 20(4):298–314

Grönroos C, Gummesson E (1985) The Nordic School of service marketing. In: Grönroos C, Gummesson E (eds) Service marketing—Nordic School perspectives. Stockholm University, Stockholm, pp 6–11

Grönroos C, Ravald A (2011) Service as business logic: implications for value creation and marketing. J Serv Man 22(1):5–22

Grönroos C, Voima P (2013) Critical service logic: making sense of value creation and co-creation. J Acad Marketing Sci 41(2):133–150

Hatami Z (2013) The role of design in service (-dominant) logic. In: Proceedings of the Naples Forum on Service, Ischia, June 18–21

Heinonen K, Strandvik T, Mickelsson KJ, Edvardsson B, Sundström E, Andersson P (2010) A customer-dominant logic of service. J Serv Man 21(4):531–548

Holm SL, Johansson U (2005) Marketing and design: rivals or partners? Des Manage Rev 16 (2):36–41

Holmlid S (2007) Interaction design and service design: Expanding a comparison of design disciplines. In: Proceedings of NorDES 2007, Stockholm, Sweden, 27–30 May

Holmlid S (2009) Participative, co-operative, emancipatory: from participatory design to service design. In: Proceedings of the 1st Nordic conference on service design and service innovation, Oslo, 24–26 November

Howells J, Tether B (2004) Innovation in services: issues at stake and trends. https://hal.archives-ouvertes.fr/halshs-01113600/document. Accessed 20 Jan 2016

Johne A, Storey C (1998) New service development: a review of the literature and annotated bibliography. Eur J Mark 32(3/4):184–252

Johnson EM (1969) Are goods and services different? An exercise in marketing theory. Unpublished dissertation, Washington University

Judd RC (1965) The structure and classification of the service market. Dissertation, University of Michigan

Kimbell L (2009) The turn to service design. In: Julier G, Moor L (eds) Design and creativity: policy, management and practice. Berg, Oxford, pp 157–173

Kimbell L (2011) Designing for service as one way of designing services. Int J Des 5(2):41–52

Kimbell L, Seidel V (eds) (2008) Designing for services—multidisciplinary perspectives. In: Proceedings from the exploratory project on designing for services in science and technology-based enterprises, Saïd Business School, University of Oxford, Oxford

Langeard E, Reggait P, Eiglier P (1986) Developing new services. In: Venkatesan M, Schmalennee DM, Marshall C (eds) Creativity in services marketing: what's new, what works, what's developing?. American Marketing Association, Chicago

Lopes LF, Godinho MM (2005) Services innovation and economic performance: an analysis at the firm level. http://www3.druid.dk/wp/20050008.pdf. Accessed 18 Dec 2015

Lovelock CH (1983) Classifying services to gain strategic marketing insights. J Mark 47:9–20

Lovelock C, Gummesson E (2004) Whither services marketing? In search of a new paradigm and fresh perspectives. J Serv Res 7(1):20–41

Mager B (2009) Service Design as an emerging field. In: Miettinen S, Koivisto M (eds) Designing services with innovative methods. University of Art and Design, Helsinki, pp 28–42

Mager B, Gais M (2009) Service design: design studieren. UTB, Stuttgart

Manzini E (2009) Service design in the age of networks and sustainability. In: Miettinen S, Koivisto M (eds) Designing services with innovative methods. University of Art and Design, Helsinki, pp 44–59

Manzini E (2011) Introduction. In: Meroni A, Sangiorgi D (eds) Design for services. Gower Publishing, Surrey, pp 1–6

Meroni A, Sangiorgi D (2011) Design for services. Gower Publishing, Surrey

Miettinen S, Valtonen A (eds) (2012) Service design with theory. Discussion on value, societal change and methods. Lapland University Press, Rovaniemi

Miles I (2010) Service innovation. In: Maglio PP (ed) Handbook of service science: research and innovations in the service economy. Springer, New York, pp 511–533

Moritz S (2005) Service design. Practical access to an evolving field. http://stefan-moritz.com/welcome/Service_Design_files/Practical%20Access%20to%20Service%20Design.pdf. Accessed 2 Feb 2016

Mulgan G (2011) Connexity: how to live in a connected world. Random House, London

Mulgan G (2015) The locust and the bee: predators and creators in capitalism's future. Princeton University Press, Princeton

Neely AD (2009) Exploring the financial consequences of the servitization of manufacturing. Oper Manage Res 2(1):103–118

Neely AD, Benedettini O, Visnjic I (2011) The servitization of manufacturing: further evidence. In: Proceedings of the 18th international annual EurOMA conference, Cambridge, 3–6 July 2011

Normann R (1982) Service management. Liber, Malmö

OECD (2000) Promoting innovation and growth in services, Organisation for Economic Cooperation and Development. OECD Publishing, Paris

OECD (2005) Promoting innovation in services. OECD Publishing, Paris

OECD (2011) Making the most of public investment in a tight fiscal environment. OECD Publishing, Paris

OECD and Eurostat (2005) Oslo Manual: guidelines for collecting and interpreting innovation data, 3rd edn. OECD Publishing, Paris

Ostrom A, Bitner MJ, Brown S, Burkhard K, Goul M, Smith-Daniels V, Demirkan H, Rabinovich E (2010) Moving forward and making a difference: research priorities for the science of service. J Serv Res 13(4):4–35

Pacenti E, Sangiorgi D (2010) Service design research pioneers: an overview of service design research developed in Italy since the '90s. Des Res J 1(1):26–33

Parasuraman A, Zeithaml VA, Berry LL (1985) A conceptual model of service quality and its implications for future research. J Mark 49:41–50

Pine JB, Gilmore JH (1999) The experience economy: work as theater and every business a stage. Harvard University Press, Cambridge

Polaine A, Løvlie L, Reason B (2013) Service design: from insight to implementation. Rosenfeld Media, New York

Prahalad CK, Ramaswamy V (2004) Co-creation experiences: the next practice in value creation. J Interact Mark 18(3):5–14

Rathmell JM (1974) Marketing in the service sector. Winthrop Publishers, Cambridge

Salter A, Tether BS (2015) Innovation in services: an overview. In: Haynes K, Grugulis I (eds) Managing services: challenges and innovation. Oxford University Press, Oxford

Sangiorgi D (2009) Building up a framework for service design research. In: Proceedings of the 8th European academy of design conference, The Robert Gordon University, Aberdeen, 1–3 April

Sangiorgi D (2012) Value co-creation in design for service. In: Miettinen S, Valtonen A (eds) Service design with theory. Lapland University Press, Vantaa, pp 97–106

Sangiorgi D, Prendiville A (eds) (2017) Designing services: key issues and new directions. Bloomsbury Academic, London

Sangiorgi D, Prendiville A, Jung J, Yu E (2015) Design for service innovation & development, final report. Available via Imagination Lancaster. http://imagination.lancs.ac.uk/sites/default/files/outcome_downloads/desid_report_2015_web.pdf. Accessed 5 Apr 2016

Schettkat R, Yocarini L (2003) The shift to services: a review of the literature. IZA Discussion Paper No. 964. Available via SSRN. http://ssrn.com/abstract=487282. Accessed 13 Oct 2015

Schneider B (2000) Portrait 9. In: Fisk RP, Grove SJ, John J (eds) Services marketing selfportraits: introspections, reflections, and glimpses from the experts. American Marketing Association, Chicago

Segelström F (2010) Visualisations in service design. Linköping University, Linköping

Shostack GL (1977) Breaking free from product marketing. J Mark 41(1):73–80

Shostack LG (1982) How to design a service. Eur J Mark 16(1):49–63

Shostack GL (1984) Designing services that deliver. Harvard Bus Rev 62(1):133–139

Stickdorn M (2010) 5 principles of service design thinking. In: Stickdorn M, Schneider J (eds) This is service design thinking. BIS Publishers, Amsterdam, pp 24–45

Stickdorn M (2011) Definitions: service design as an interdisciplinary approach. In: Stickdorn M, Schneider J (eds) This is service design thinking. BIS Publishers, Amsterdam, p 373

Stickdorn M, Schneider J (2010) This is service design thinking. BIS Publishers, Amsterdam

Stigliani I, Fayard AL (2010) Designing new customer experiences: a study of socio-material practices in service design. Imperial College Business School, London

Swartz TA, Bowen DE, Brown SW (1992) Fifteen years after breaking free: services then, now, and beyond. Adv Serv Mark Man 1:1–21

Tether BS (2003) The sources and aims of innovation in services: variety between and within sectors. Econ Innov New Technol 12(6):481–505

Tether BS (2013) Services, innovation, and managing service innovation. In: Dodgson M, Gann D, Phillips N (eds) The Oxford handbook of innovation management. Oxford University Press, Oxford

Thackara J (2005) In the bubble: designing in a complex world. MIT Press, Cambridge

Trott P (2012) Innovation management and new product development, 5th edn. Pearson Education, Essex

Van Ark B, Broersma L, den Hertog P (2003) Services innovation, performance and policy: a review. Synthesis report in the framework of the project on Structural Information Provision on Innovation in Services (SIID). University of Groningen and DIALOGIC, Groningen, NL

Vandermerwe S, Rada J (1988) Servitization of business: adding value by adding services. Eur Manage J 6:314–324

Vargo SL, Lusch RF (2004) Evolving to a new dominant logic for marketing. J Mark 68(1):1–17

Vargo SL, Lusch RF (2008) Service-dominant logic: continuing the evolution. J Acad Mark Sci 36 (1):1–10

Vargo SL, Lusch RF (2014) Service-dominant logic: premises, perspectives, possibilities. Cambridge University Press, Cambridge

Vargo SL, Lusch RF (2016) Institutions and axioms: an extension and update of service-dominant logic. J Acad Mark Sci 44(1):5–23

Wetter-Edman K (2011) Service design: a conceptualization of an emerging practice. University of Gothenburg, Gothenburg

Wetter-Edman K (2014) Design for service. A framework for articulating designers' contribution as interpreter of users' experience. University of Gothenburg, Gothenburg

Wetter-Edman K, Sangiorgi D, Edvardsson B, Holmlid S, Grönroos C, Mattelmäki T (2014) Design for value co-creation: exploring synergies between design for service and service logic. Serv Sci 6(2):106–121

Wilson A (1972) The marketing of professional services. McGraw-Hill, London

Zeithaml V, Parasuraman A, Berry L (1985) Problems and strategies in services marketing. J Mark 49:33–46

Chapter 3
How to (Re)Design Services: From Ideation to Evaluation

Abstract This chapter explores service design as a multidisciplinary, holistic and people-centered approach that can support public and business organization in the development of new services or the redesign of existing services that answer to contemporary innovation requirements. Contributions from the academic and the professional fields are analyzed to describe the core characteristics of the approach and the key steps of the service design process. From the capacity of understanding people, contexts and relationships to the definition of scenarios and concepts, from the development and validation of service ideas to their implementation, each step of the process is explored in depth, including a description of its general purpose, activities to be conducted, and suggestions about useful tools typically adopted in service design practice. Moreover, concerning the development of service ideas, a particular focus on what needs to be designed in services is provided. To conclude, the topic of evaluation is introduced as the missing element in the service design process to give solidity and reliability to service design interventions, to prove the value of solutions proposed, and demonstrate the benefits of adopting service design itself.

Keywords Service design · Service redesign · Service design approach · Service design process · Service design tools · User research · Service development · Prototyping · Service implementation

3.1 Designing New Services and Service Redesign

Service innovation is usually described in relation to the development of new services, which includes the introduction of a new service concept, offering, and delivery system, or to changes in the offering and delivery system of existing services (see Sect. 2.4). This is also asserted in studies on *New Service*

Development (NSD),[1] where a *new service* is defined as both an offering not previously available to customers that implies a radical innovation or transformational change (Chapman 2002), or incremental changes to an existing offering that customers perceive as being new (Johnson et al. 2000). In both cases, it must also be considered that nowadays service innovation can be strictly connected to the emerging and adoption of new technologies and technology-enabled actions, affecting ways of experiencing the service or organizing back-office activities. Smart cards, for example, represent a simple but disruptive application of technology that has enabled the redesign of many traditional services.

Even though service design is often referred to as an approach that can address both the creation of new services and service redesign, most of the literature mainly focuses on how service design can enable the development of new service solutions rather than improving existing ones (Berry and Lampo 2000). This makes us wonder about the differences (if there are any) between designing and redesigning a service, and if it is necessary to face them differently regarding the service design process, competencies, and resources involved.

According to NSD studies, within the development of new service offerings, service design represents the stage dedicated to the definition of the prerequisites necessary to achieve service quality, consisting of the service concept, system and process (Edvardsson 1997), thus remaining excluded from the phases aimed at service implementation. On the contrary, Yu and Sangiorgi (2014: 195) describe service design as "an *approach* or *thinking* that can be transferred to a wide variety of practices for service innovation," including NSD rather than being simply one of its phases. However, they also recognize that service design lacks the capabilities to contribute to service implementation. This strengthens the need to further explore the contribution of service design to service development, and as a consequence to service innovation, in order to better frame capabilities and competencies required, articulate the approach according to the context of application and the type of project (creation of a new service or redesign), and finally confirm its legitimacy (Stigliani and Tether 2011). To do so, according to Mulgan (2014), service design should pay more attention to organizational issues and cost-effectiveness of the solutions proposed. In our opinion, it is also important to start measuring the effects generated by service design interventions on organizations, users, and the society at large.

The question is: how does this apply to the creation of new services rather than to service redesign? In the face of *app-centered* design solutions that bring to the launch of hundreds of service start-ups that quickly fail, the literature suggests, as stated before, that service firms can innovate with what exists as well as with what still needs to be created. What changes when facing redesign rather than the

---

[1]NSD can be considered the service-equivalent of New Product Development, the process of introducing a new product in the market. NSD focuses on the practices necessary to design a new service (Johnson et al. 2000: 3), where service design represents a specific phase characterized by a set of activities, tools and competences (Edvardsson 1997; Johnson et al. 2000; Sangiorgi et al. 2015; Yu and Sangiorgi 2014).

development of new services is that redesign implies that organizations assess the way services are delivered and received, and create more effective ways to serve their users (Berry and Lampo 2000: 266). This requires changing the initial step of the service design process from the analysis of the context and the understanding of user needs to the evaluation of the service to be improved.

From a service design point of view, this difference between the design of new services and service redesign, in addition to the need for evidence about the contribution of service design to service development and innovation, introduces a reflection about an evaluative dimension that should be included in the service design practice. To fully understand how to address this issue, that in our opinion represents a necessary evolution of the discipline, we firstly need to explore the current characteristics attributed to the service design approach, as well as the acknowledged steps of a traditional service design process.

3.2 The Service Design Approach

As stated in Sect. 2.5, service design is a multidisciplinary practice that requires collaboration between different competencies and that entails a process made of various stages, from research to concept development, and also considering, to a certain extent, service implementation. The objects of service design are complex systems that involve both physical and intangible elements, such as interactions and processes among different players. Within this framework, adopting a service design approach means understanding or creating the conditions for service development by taking into consideration all the components that characterize the service system.

The service design approach has been described in several ways since the introduction of the term (see Sect. 2.5), both in the academic and professional fields. In the academic field, just to mention a few, Manzini (1993) described service design as an activity that links the techno-productive dimension to social and cultural aspects. Accordingly, Morelli (2002) described service design as a domain of knowledge entailing the organizational and design culture, and the social construction of technology. In her doctoral research, Sangiorgi (2004) explored the service design approach through the *Activity Theory*, considering services as a particular type of human activity and focusing on the role of *service encounters* as situated actions (Sangiorgi and Maffei 2006) shaped by the sociocultural context of the user and the organizational environment. In general terms, we can affirm that scholars consider service design as an approach endowed with the capability to build a relation between a context, an organization, and people through the realization of a service performance.

On the other hand, in the professional field, service design is mainly described as an activity endowed with the potential of generating value for the organization granted by the adoption of a user-centered approach. As asserted by Mager and Sung (2011: 1),

Service design aims at designing services that are useful, usable and desirable from the user perspective, and efficient, effective and different from the provider perspective. It is a strategic approach that helps providers to develop a clear strategic positioning for their service offerings. Services are systems that involve many different influential factors, so service design takes a holistic approach in order to get an understanding of the system and the different actors within the system.

Because of its continuous evolution and the interplay of several disciplines, service design is described in many other ways. Nisula (2012) analyzed definitions provided by service marketing, management and design scholars from 1999 to 2011. What emerges from her study is that some definitions are very broad and not focused, and service design seems to be capable of taking care of the whole service development and production process; while other definitions are more compact and focused on the human experience or single aspects of the service solution, like touchpoints. All things considered, it is quite obvious that a commonly accepted vision of the service design approach, and its capabilities and characteristics, still needs to be found. Not by chance, Stickdorn and Schneider (2010: 29) affirmed, "if you would ask ten people what Service Design is, you would end up with eleven different answers—at least". However, some similarities can be traced and allow us to imagine new directions for the service design practice:

- Service design is a holistic, user-centered approach focused on the relation between provider and user;
- The service user is at the center of the experience over time, i.e. before, during and after the effective use of the service;
- The user experience is made available by actors, processes and activities provided by or connected to the service provider;
- The application of service design competencies within an organization can result in solutions that bring increased user satisfaction, more compelling brands, and the acceleration of new ideas to market, establishing improved or new processes for service creation and development that more effectively support innovation.

Once having explored the key features of the service design approach we can focus on how to actually design (or redesign) services, by analyzing the steps of the service design process and the activities they imply.

3.3 The Service Design Process: From Research to Implementation

The conceptualization of the service design process has been (and still is) widely discussed. Both in the academic and the professional fields many representations have been proposed, typically made up from three to seven or more steps. However, they ultimately look very similar (Stickdorn and Schneider 2010) and share the same idea about an initial step focused on exploration and research up to a final step of delivery of the solution to be implemented and launched on the market. The most acknowledged model is the famous 'Double Diamond' developed by the British Design Council,[2] which is structured in 4 phases (discover, define, develop and deliver) alternating divergent and convergent moments in an iterative process. Based on the 'Double Diamond', in the book *This is Service Design Thinking*, Stickdorn and Schneider (2010) describe a service design process made of four phases: exploration, creation, reflection and implementation. They emphasize the iterative nature of every step, which may require restarting from scratch, and the necessity to begin the process by designing the process itself according to the context of the project. However, even though all existing frameworks share the same principles (despite wording differently), they recognize this is "a very basic approach to structure such a complex design process" (Stickdorn and Schneider 2010: 126).

Meroni and Sangiorgi (2011) identify as well four steps in the service design process: analyzing, generating, developing and prototyping. In this case also, the process begins with a research phase, but differently to the 'Double Diamond' it closes with a prototyping phase aimed at the materialization of the service concept. As a consequence, while the *develop phase* of the 'Double Diamond' is aimed at the creation, prototype and test of concepts through several iterations up to the selection and delivery of the final solution, the *developing phase* proposed by Meroni and Sangiorgi focuses on the elaboration of service ideas into more detailed aspects, thus excluding from the process the implementation phase. To mention other examples, Moritz (2005) outlines six phases: understanding, thinking, generating, filtering, explaining, realizing. Van Oosterom (2009) describes the service design process composed of five steps that consist of discovering, concepting, designing, building and implementing. And so on.

In the professional field, it is very common for service design agencies to communicate their approach by visualizing the service design process they follow and the tools they adopt. We describe some of them to exemplify the overall panorama. IDEO, one of the most famous international design agencies, proposes a unique process for the design of product and services, which is made of 5 steps: observation, synthesis, idea generation, refinement and implementation. This is the result of a wider vision that IDEO advocates all over the world and that has acquired a great relevance in service design practice, so-called *design thinking*. In the words of Tim Brown, president and CEO of IDEO, *design thinking* is a human-centered approach to innovation that can support organizations in developing better products, services, processes, and strategy by thinking like a designer, that is, understanding the human point of view together with what is technologically feasible and economically viable. Adopting this approach non-designers are enabled to use creative tools and become active actors in the design process. Being inherently user-centered, service design if often compared or assimilated to *design*

---

[2]See for example: https://connect.innovateuk.org/documents/3338201/3753639/Design+methods +for+developing+services.pdf.

thinking and human-centered design. However, even though the design process is the same, we must not forget that service design is not just about generating creative ideas and bringing a lateral vision, but rather a system-oriented approach aimed at producing valuable solutions both from the user and the provider organization. Moreover, differently to *design thinking* and human-centered design that (ideally) apply to any project, service design focuses on services and related processes, experiences and interactions, adopting service-specific design tools (e.g. the *system map* and the *user journey map*) to describe them.

Another service design agency, Live|Work, proposes a service design process made of six steps: understand, imagine, design, create, enable and improve. Similarly to the others, this process includes: the understanding of the experience of customers to frame the challenge posed by the organization; a concept generation phase (imagine) supported by the use of creative methods; a prototype phase (design) aimed at detailing and validating the concept; and a delivery phase (create) where the solution is finally implemented. Live|Work then adds two more phases to the service design process that in most cases are not explicitly expressed: an *enabling phase* with the function of supporting the organization in the adoption of the new solution; and an *improvement phase* as an ongoing activity that should accompany the evolution of the service according to market, technical, regulatory and other changes in order to maintain a good level of service quality over time. This is done through customer experience tools and metrics, introducing an explicit evaluative dimension into the service design process, which usually is implicitly and exclusively associated with the selection and validation of ideas in the prototype phase.

Engine also, a pioneer agency in the field, proposes a measurement step in the description of its approach. In this case, the service design process is made of three phases (identify, build and measure), further detailed through eight sub-activities (orientate, discover, generate, synthesize, model, specify, produce, measure) that circularly follow one another. The evaluation dimension is here made even more explicit and refers to the idea of being able to measure what customers and providers value (in terms of efficiency and effectiveness, but also desirability, usefulness and usability of services) to validate the designed solution on one hand, and to inform ongoing improvements on the other hand.

In most cases, the description of the service design process is often supported by a toolkit where service design and *design thinking* tools are proposed in order to accompany practitioners (and eventually a wider audience of non-designers operating in client organizations) in dealing with service design projects and educating at the approach itself. Toolkits may include specific tools for conducting user research, generating and visualizing ideas, detail and prototype solutions, and so on, such as those proposed by IDEO,[3] Live|Work[4] and Namahn in collaboration with

---

[3] www.ideo.com/tools.

[4] www.liveworkstudio.com/tools.

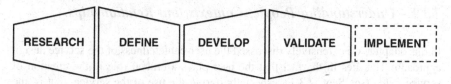

Fig. 3.1 The service design process: from research to implementation

Flanders DC.[5] They can be focused on tools that facilitate the organization of design activities, such as the 'Design Sprint Kit' developed by Google.[6] Or they can be dedicated to specific service categories, such as the practical guide 'Designing for public services' written by Dahl et al. (2016) on behalf of Nesta Foundation[7] and IDEO with the support of the 'Design for Europe'[8] program. Moreover, toolkits often include tools for the organization of co-design sessions with users, employees, and other stakeholders of the client organization.

This variety of processes and tools gives evidence to the fact that service design is becoming increasingly acknowledged in several fields of application, and that several design agencies are expanding their businesses thanks to the inclusion of these kinds of competencies. On the other hand, even though there are wide similarities among their approaches, especially concerning the importance of understanding the user point of view (both in the beginning and in the end of the process to validate solutions) , a univocal vision about when, where and how service design competencies should intervene appears to be lacking once again. In particular, the role and the contribution of service design to the implementation phase are not fully clarified in the processes described above, and evaluation is mentioned only in few cases and without a clear objective.

Based on knowledge gathered and considerations discussed in this section, we explore as follows those phases of the process that in our opinion are necessary to face a service design (or redesign) project (Fig. 3.1). Since in most of the cases they are described very shortly, and service design literature only superficially explores how they can be pragmatically conducted, and what are the elements to be designed besides the user experience, the contents of the following sections also rely on our experience as service design researchers and teachers. The purpose is not to propose a new version of the service design process, but rather to develop a common vision that could unify existing ones, to reflect in the end on the role and the purpose evaluation could have in some steps.

---

[5] www.servicedesigntoolkit.org.
[6] https://designsprintkit.withgoogle.com/planning/.
[7] www.nesta.org.uk.
[8] www.designforeurope.eu.

3.3.1 Understanding People, Contexts and Relationships

In this chapter, we extensively discussed the user-centeredness of the service design approach. This lies in the fact that the user plays a crucial role in the creation of service value (see Sect. 2.3.1), especially during the use of the service, that is, the moment during which the user directly interacts with the service (provider). *Using* or *consuming* the service means experiencing the service (Grove and Fisk 1992), and in particular the service offering delivered by the service provider (often thanks to a system of suppliers and other stakeholders) through those tangible elements of the service that are defined as *touchpoints*. Thus, designing a service requires understanding what the needs for a service experience to exist in the mind of the user are, and what characteristics it needs to have to meet his/her expectations, perceiving it positively and keeping using the service. This is why research is a fundamental ingredient in the service design approach, and it should be conducted as the first step of the service design process.

The research that is typically conducted in service design practice is mainly qualitative (Polaine et al. 2013), and because of the experiential nature of services, it is often based on observational techniques (Grove and Fisk 1992). The purpose is to explore user behaviors during everyday life, emotions and motivations that bring people to opt for a solution, and make a choice rather than another one. This does not mean that quantitative research is not useful in service design, but as Polaine et al. (2013) suggest, service designers need to know the underlying factors that make people behave in a given way to tackle a design problem correctly. Knowing how many people act in a specific way may be useful as well, but with the support of qualitative insights that justify these numbers. On the other hand, numbers become fundamental when research investigates the market at a wider scale, and the provider-side of the service (costs and revenues, resources expenditures, process flows, etc.), even though, also in these cases, qualitative insights might be relevant. We refer for example to research based on market trends and innovation drivers (through a desk analysis of trend reports typically developed by big consulting companies[9] as result of their interpretation of contemporary phenomena and annual practices), or the collection of best practices and case studies as benchmarking methodology. This kind of desk research can help frame the context of the project at a socio-economical level to drive design choices that are coherent with emerging needs of the society, and new business models characterizing the market. Qualitative research concerning the provider organization could then coincide with user research but applied to employees, managers and other actors involved to investigate the needs, expectations and perceptions of the organization as a whole, and concerning the service offered and related delivery processes and activities. To

---

[9]In the design field, see for example the trend report proposed every year by Fjord https://trends.fjordnet.com/trends/. Outside of the design field, interesting studies are developed by Price Water Coopers (PWC) or can be found on online platforms such as trendwatching.com and springwise.com.

sum up, the research phase should ideally include both quantitative and qualitative data, and the analysis should investigate: user needs, expectations, and perceptions, the production context (the provider-side of the service), the target market, the socio-economical context.

In our opinion, a good research strategy could consist of initial data crunching of quantitative data about the market and the organization (and eventually its competitors). This aims at identifying market directions, behavioral and consumption patterns, and all the information useful to define the state of the art of the context in which the project is situated, as well as design drivers that could inform the resolution of the design problem. Thanks to the diffusion of open and big data, nowadays also service designers can more easily access this kind of information to give solidity to solutions they propose (Foulonneau et al. 2014). The analysis should then move to the collection of qualitative insights through user research methods (also applied to the provider context), observing people in the specific service context.

There is a broad range of service design tools aimed at exploring the environment in which the design takes place: some of them have been specifically developed for the service design practice, while others come from other disciplines, such as ethnography and marketing (Moritz 2005). Moreover, a distinction needs to be made between tools aimed at data collection, and tools aimed at data visualization and interpretation. In fact, a key element distinguishing design tools from other methods and techniques is the role of the visual. Visualization skills of service designers represent an added value for data interpretation because they capture the complexity of services and customers in more simplified formats (Segelström 2009).

Concerning data collection, service design practitioners usually adopt standard ethnographic methods such as *focus groups, interviews, direct* or *participant observations, mystery shopping,* or they opt for more sophisticated (design) tools such as *shadowing, card sorting, diaries, service safari,* and *cultural probes.*[10] Some of them allow for a close observation of single individuals; others can focus on behaviors and interactions within a community. In some cases, the use of tools requires a researcher or designer to guide the observation, while in other cases (e.g. *cultural probes*) the user is trained in the utilization of the tool and conducts a self-observation.

Other tools can then enable data interpretation. Widespread ways to interpret research insights about potential service users consist for example of the creation of *personas,* user archetypes that gather characteristics, behaviors and attitudes of people observed. The visualization of research results through conceptual maps is also very common in service design practice: they can focus on actors and stakeholders, highlighting roles, relationships and back-office activities, or concentrate on touchpoints and the user journey. A good solution often used at the beginning of

---

[10]We do not provide a description of tools in this book, since we refer to very common techniques that have widely been explored by literature and that can be easily found online.

a service (re)design process to build a complete knowledge of the service, including both the user experience and back-office activities, consists in compiling the so-called *service blueprint*. As said above, interpretation of qualitative data should then be completed by quantitative information. *Surveys* and *questionnaires* (that can be administered offline or online) are for sure the quickest and easiest way to reach large numbers of people. In this case, interpretation is usually enabled by the visualization of results through graphs and infographics.

To conclude, understanding people, contexts, and relationships helps define the overall framework that is necessary to trigger the following concept generation phase, starting from the definition of scenarios that are based on people and organization's needs, expectations, behaviors, and so on, but also on unexpected insights emerging from research. The next section illustrates what in our vision is the second step of the service design process.

3.3.2   Defining Service Scenarios and Concepts

This phase of the service design process is based on divergent thinking aimed at transforming research insights and interpretations into ideas. At this stage, the involvement of users and other actors from the provider organization is crucial to develop solutions that bring value at all levels of the service system through the orchestration of back-office processes, experiences, and related touchpoints (including physical and digital channels) in a coherent way. Designing with people rather than for people, making them active players in the design process, is an essential characteristic of the service design approach that in this phase becomes explicit. In fact, according to Steen et al. (2011), the benefits of co-design in service design practices are identified as:

- Improving the creative process and organization of the service project;
- Better matching offer and needs from a user perspective;
- Enabling creativity, awareness of customers, and internal cooperation on innovation from a provider perspective.

Several ideation techniques, mainly borrowed from *design thinking*, can be used to inspire brainstorming sessions (Stickdorn and Schneider 2010), and supporting people (including non-designers) in visualizing their ideas. We could mention for example *mind mapping* or the *Disney method*.[11] Service design toolkits cited in Sect. 3.3 offer a broad range of solutions to facilitate co-design sessions. Results of the research phase visualized for example through *personas*, *user journey maps* and *blueprints* are also used to stimulate group discussions. Moreover, an important role

---

[11]The 'Disney Method' is a brainstorming technique developed by Robert Dilts in 1994. It is based on parallel thinking and aims at bridging the gap between dream and reality, turning unexpected ideas into solutions for existing problems.

could be played by *service scenarios*. *Service scenarios* are visual or textual stories telling how people will interact with a new service (or a service context) in the next future. They can be represented through texts, storyboards or even videos, and can be used at this stage of the process or later to validate the service idea. At this stage, *service scenarios* can be both the result of an ideation session, or can kick-start the session itself. They can be posed in negative terms, highlighting problematic aspects of a situation or a solution to focus on opportunity areas for service design. A simple way to prompt the definition of service scenarios is asking the question *what if...?*, which helps to imagine a future change enabled by the introduction of a given service experience (e.g. what would happen if shopping at the supermarket could be done at home?).

In general, ideation sessions aim at generating a huge number of ideas at a very high concept level. A selection process, which can be part of the current or future co-design session as well, is then set up to identify the most promising concepts up to the selection of the solution that will be developed in the following steps of the process.

3.3.3   Developing Service Ideas

Developing a service idea means bringing the service concept at a deeper level of detail. It entails the design of all the elements of the service system that are necessary to the provider to deliver the service, and all the elements of the service experience that are needed for the user to experience the service. These elements consist of performances and relations occurring within the system (Glushko and Tabas 2009), which means designing processes operated by actors inside and outside the provider organization, and the relations between them that are necessary to perform the service offering, and that support the business model of the service. This implies taking into consideration resources, competencies, and technologies required, and changes that will affect them according to the evolution of the context.

On the other hand, referring to the service experience, what needs to be designed are the service offering, interactions between the user and the service, and touchpoints that allow these interactions to happen. Developing the service offering does not only depend on user needs and expectations but also concerns the positioning of the service in the market in relation to competitors and similar solutions (Johnson et al. 2000). The service offering is strictly connected (but slightly different) to the *value proposition*, which Grönroos and Voima define (2013: 145) as "a promise that customers can extract some value from an offering". It could be said that the service offering is a medium through which users can achieve the service promise, and is made of what the user can buy or get access to in concrete terms. For example, thinking about a hotel service, the offering could consists of the different types of rooms and additional services (such as breakfast, spa, etc.) the user can access; while the *value proposition* is given by values associated with the offering (comfortable spaces, romantic atmosphere, good position, etc.). As said above, the offering is also connected to several activities, both back-office and front-office (Glushko and Tabas 2009), that enable the user accessing it (for example, booking, check-in, customer service, etc.), and that should be designed accordingly as part of the service experience (and the service performance from the organization point of view). Moreover, all these activities involve some interactions between people, people and spaces, and people and objects that constitute the service touchpoints (Bitner et al. 1990). Interactions, as well as touchpoints, can be physical or digital (such as online platforms and digital interfaces), and also include information necessary to the user to know and use the service (Clatworthy 2011).

These elements also contribute to the achievement of the *value proposition* and need to be designed coherently. The definition of the overall *value proposition* and values that should characterize every service element is fundamental firstly to validate the idea developed, and then to evaluate the service once it has been launched on the market (see this chapter). For example, if the solution proposes a quick booking process as one of the values that must characterize the service experience, it will be necessary to evaluate the level of speed of the booking before implementing the service, and monitor that the same level is maintained (or needs to be changed because of changing user expectations) once the service is implemented.

To design such a complexity, service design practitioners use several tools that, in this case also, mainly consists of visual maps. The most important are:

- The *offering map*, which is useful to define the primary and secondary offering (i.e. the core of the service offering and additional or specific services) (Hoffman and Bateson 2010) and connect it to the touchpoints;
- The *system map*, which visualizes actors involved in the service system (internal staff, external suppliers, partners, investors, etc.) and their relations (regarding functions and resources exchanged);
- The *service blueprint*, which, differently from when used as a research tool, allows here detailing back-office and front-office processes in all the steps of the service experience, also highlighting interactions and touchpoints;
- The *customer journey map*, which focuses exclusively on the user experience, highlighting step by step all the options available to the user. The user experience is often visualized through storyboards and videos as well.

Once the solution is developed (on paper), it needs to be validated before becoming a real service available on the market. The process of validation represents in our opinion the next step of the service design process.

3.3.4   *Validating Service Ideas*

Once the service is defined and developed on paper, it is necessary to transform it into a real experience. This means validating all the elements individually, as well as the overall system and user experience through an iterative process with different levels of depth. Service ideas can be validated at an early stage of the project or later, once the service is fully developed and ready to be implemented (Blomkvist 2012; Coughlan et al. 2007). Preliminary concepts are usually prioritized, selected, and tested through *rapid prototypes* of the user experience and the most relevant service interfaces, to investigate the crucial interactions occurring between the service and its final users. While during the development phase, validation entails the full service experience, including physical and digital touchpoints represented through detailed prototypes that are as close as possible to the final solution. Addressing this activity appropriately is fundamental to reduce the risks to lose competitive advantage, and to waste resources in the implementation of a poor service (Rae 2007).

The shift from the design of tangible elements to experiences, interactive systems, and services enabled the creation and adoption of new prototyping methods able to visualize interactions between people, places, technology, and environments (Buchenau and Fulton Suri 2000). Prototyping represents a core activity for the service designer to explore, evaluate, or communicate service solutions (Blomkvist 2014; Holmlid and Evenson 2008; Wetter-Edman 2011). Prototypes can range from low-fidelity outputs to high-fidelity models and in service design practice can apply to processes, interactions, and artifacts (Passera et al. 2012). Nesta (2011), referring to the public sector, identifies four elements to be prototyped within service design projects: physical elements, the system of structures and processes, information and roles, skills and behaviors.

It is also possible to outline different purposes for prototyping activities. They have an explorative purpose when prototypes (usually quick and rough) are made to inspire ideas and solutions (Blomkvist 2014). They have an evaluative purpose when, through prototyping, alternative design solutions are evaluated before a final design solution is reached (Weick 2001). Referring to evaluative prototypes, measuring how people and organizations perceive the prototype of a service is as difficult as measuring the service itself. The difference may consist in the fact that service prototypes only represent some touchpoints or processes, thus making the design of the evaluation strategy restricted to a smaller context. The validation process can also be conducted in a collaborative manner (Coughlan et al. 2007), involving end users and different competencies and roles within the organization (i.e. marketing, communication, and management staff).

Different techniques can be adopted for the development of rapid and more complex prototypes. The most used ones to prototype the user experience are *role playing, scene enactment* and *experience prototyping* (Miettinen 2009). These allow designers validating processes and interactions happening at specific moments of the service experience, as well as the overall customer journey, representing user–provider relationships and highlighting process inconsistencies. Regarding physical elements, such as spaces, objects or interfaces, prototypes usually consist of *mockups*, *wireframes*, and *beta-versions* (of digital components). Once the validation process is concluded, the solution is refined and ready to be implemented.

3.3.5   Supporting the Service Implementation

The transition from prototyping to implementation is a very delicate phase, which requires the setting up of systems or the introduction of changes within the organization, concerning processes, people, procedures, as well as technical systems. It entails the idea of making concepts real in terms of design, economic, technological, and operational aspects, including the personnel, the structures, and the delivery activities (Lovelock and Wirtz 2011; Shostack 1984). In this phase, services are launched on the market in a full or partial version. First of all, this means planning the *service roadmap*, that is the various service implementation steps over time. Once the roadmap is defined, the provider needs to set up the service system, establishing ownership and responsibilities, activating the system of partners and suppliers, and the related channels, and structuring back-office activities necessary to the service delivery. To do so, the organization needs to operationalize front-office processes necessary to deliver the service offering, defining prices, and the communication system, as well as realizing the touchpoint system that enables service interactions. Finally, the service can be performed and experienced.

In some cases, mainly referring to digital services, the implementation phase can aim at producing a hi-fidelity simulation that is launched on the market before the full service is actually implemented. This is the case of *beta-versions* that quickly turn into *perpetual beta*. Digital services are usually delivered on a global scale so that newer versions are immediately available to all users. This makes the service a sort of perpetual prototype, where improvements become unnoticed to the market, and the innovation process consists of a continuous evaluation–improvement cycle. It is not possible to consider *perpetual beta* for non-digital services, where physical facilities need to be upgraded and personnel to be trained (Ahson and Ilyas 2011).

The role of service design at this stage of the process remains underexplored (Kimbell 2011). Few organizations have developed internal service design teams, but in this situation also, decision-making concerning service implementation is often driven by non-designers (Tether 2008). While service designers guide the earlier phases of the process, implementation is almost exclusively delegated to the organization and led by product managers, product development departments, and customer departments (Lee 2016).

3.4   Including Evaluation in the Service Design Process

Based on the exploration of the service design approach and process, we conclude this chapter reflecting on a new element, evaluation, which so far has been barely treated in service design literature and practice as a consequence of the need for the discipline to firstly determine itself and its boundaries.

Services have come to dominate our economies. Whether you manage a traditional service firm or a manufacturing company, adding value through services has become an essential way to compete. [...] Today customers are looking for service value, comprehensive solutions, and memorable experiences. (Gustafsson and Johnson 2003: xiii)

This quote by Gustafsson and Johnson well summarizes what has been discussed until now. On the one hand, services have come to dominate our economies, opening up a broad debate about their nature, their value for people and organizations, the concept of service innovation, the necessity to establish dedicated disciplines, and develop new skills to study and deal with these issues. On the other hand, service design, as one of the disciplines emerged in response to the increasing importance of services, has struggled to be acknowledged as a driver of innovation concerning the development of valuable services that answer to continuously changing requirements, on the economic, environmental, technological and social level. The discourse on the actual contribution of adopting the service design approach (and how to measure it) is constantly evolving, as well as how it can effectively deliver solutions that could become services that answer to the quest for service value (and innovation in a broader perspective).

Contemporary requirements are pushing service design toward new challenges that require stepping back from the creative process to include a wider range of strategic and management activities (Kimbell 2009). Creating successful services by paying attention to both the needs of the provider and the user experience, as well as addressing the complex challenges of our society, is the foundation that service designers need to build over the next decade to inform, drive, and enable change within organizations (Sangiorgi et al. 2015). However, despite the service design approach and process having been extensively conceptualized, as demonstrated in this chapter, and several tools adopted, adapted or created for the specific purpose of designing services, the lack of rigorous theory and principles are still perceived (Kimbell 2011; Sangiorgi 2009). This may lie in the fact that, while there is a clear and (mostly) unanimous vision about designing for and with the user, less attention has been given so far to designing for the organization that provides the service. This is demonstrated by the fact that in the service design process no distinction is made between designing a new service and redesigning an existing service. Accordingly, the implementation step is rarely explored and explained in literature and the approaches proposed by service design agencies. Another reason could be that, when designing a service, no evidence is provided about the value of solutions developed, either before or after they have been implemented.

User observations, case studies, and narrations typically used by service designers are undoubtedly useful to understand and describe changes in the market and the society, user needs, and expectations. But even the most accurate user analysis, as well as any service built upon it, represents a subjective and probably delimited case, where no understanding is built upon the dynamics that lead to success or failure, and to innovation or breakdown. Both service providers and designers need to prove and learn how to improve the value of the services they deliver, going beyond the correctness of procedures, the introduction of technological progress, and the satisfaction of emerging user needs, but evaluating all of them in the same systemic perspective that is typical of service design. Similarly, a systematic way to measure the effect of adopting the service design approach on the organization has not yet been conceptualized. Considering service design as a form of transformation design, Burns et al. (2006) argued that

> One of the biggest challenges practitioners are facing is about communicating the value and impact of a transformation design process. As journalist Geraldine Bedell put it, 'It's difficult to get a handle on this stuff. You can photograph a new car for a magazine; you can't photograph new traffic flows through a city. So that's one reason why there's so much suspicion'. Stakeholders who have participated in transformation design projects are enthusiastic champions of the work. But in order to inspire those at a company board or ministerial level, we need to build up an appropriate shared language and evidence base. (Burns et al. 2006: 27)

More recently, some service design scholars and practitioners (see e.g. Blomkvist 2011; Foglieni and Villari 2015; Løvlie et al. 2008; Manschot and Sleeswijk Visser 2011; Polaine et al. 2013) have started reflecting on the role of evaluation in service design practice, and for determining the value of service design, but contributions in this sense are still rare and fragmented. What seems to be acknowledged is that understanding how to determine the value of services resulting from service design interventions could contribute in bringing evidence about the value of service design itself.

Thus, how could we measure if a service design solution is going to become a successful service? And, in the case of redesign, how could we evaluate an existing service to develop appropriate redesign solutions? In the *Design Dictionary* published by Birkhäuser, Mager (2008: 355) provides the following definition of service design:

> Service design addresses the functionality and form of services from the perspective of clients. It aims to ensure that service interfaces are useful, usable and desirable from the client's point of view and effective, efficient and distinctive from the supplier's point of view.

Regarding the competencies required to service designers, she continues asserting that

> Service designers visualize, formulate and choreograph solutions to problems that do not necessarily exist today; they observe and interpret requirements and behavioral patterns and transform them into possible future services. This process applies explorative, generative and evaluative design approaches, and the restructuring of existing services is as much a challenge in service design as the development of innovative new services.

In this definition, an evaluative and critical competence is explicitly assigned to the service designer, in support of the creative one, as well as the purpose for service design to deliver solutions that are endowed with specific values that address both the user and the provider organization. Our goal in this book is to make this evaluative component more explicit in the service design process and provide a hypothesis for its operationalization. To do so, in the next chapter, we explore some interesting cases and experts' opinions that bring useful insights to frame the topic.

References

Ahson SA, Ilyas M (eds) (2011) Service delivery platforms: developing and deploying converged multimedia services. CRC Press, Boca Raton

Berry LL, Lampo SK (2000) Teaching an old service new tricks: the promise of service redesign. J Serv Res 2(3):265–275

Bitner MJ, Booms BH, Tetreault MS (1990) The service encounter: diagnosing favorable and unfavorable incidents. J Mark 54:71–84

Blomkvist J (2011) Conceptualising prototypes in service design. Linköping University, Linköping

Blomkvist J (2012) Conceptualisations of service prototyping: service sketches, walkthroughs and live service prototypes. In: Miettinen S, Valtonen A (eds) Service design with theory: discussions on change, value and methods. Lapland University Press, Vantaa, pp 177–188

Blomkvist J (2014) Representing future situations of service: prototyping in service design. Linköping University Press, Linköping

Buchenau M, Fulton Suri J (2000) Experience prototyping. In: *Proceedings of the 3rd Conference on Designing interactive systems: processes, practices, methods, and techniques*, New York, 17–19 Aug 2000

Burns C, Cottam H, Vanstone C, Winhall J (2006) RED paper 02: transformation design. Available via Design Council. http://www.designcouncil.org.uk/resources/report/red-paper-02-transformation-design. Accessed 10 Sept 2016

Chapman JA (2002) A framework for transformational change in organisations. Leadersh Org Dev J 23(1):16–25

Clatworthy S (2011) Service innovation through touch-points: development of an innovation toolkit for the first stages of new service development. Int J Des 5(2):15–28

Coughlan P, Fulton Suri J, Canales K (2007) Prototypes as (design) tools for behavioral and organizational change: a design-based approach to help organizations change work behaviors. J Appl Behav Sci 43(1):122–134

Dahl S, Roberts I, Duggan K (2016). Designing for public services. Available via Nesta. http://www.nesta.org.uk/sites/default/files/nesta_ideo_guide_jan2017.pdf. Accessed 4 Apr 2017

Edvardsson B (1997) Quality in new service development: key concepts and a frame of reference. Int J Prod Econ 52:31–46

Foglieni F, Villari B (2015) Towards a service evaluation culture. A contemporary issue for service design. In: Proceedings of Cumulus conference, Milan, 3–7 June 2015

Foulonneau M, Martin S, Turki S (2014) How open data are turned into services? Exploring services science. In: Snene M, Leonard M (eds) Exploring Services Science. IESS 2014. Lecture Notes in Business Information Processing, vol 169. Springer, Cham, p 39

Glushko RJ, Tabas L (2009) Designing service systems by bridging the "front stage" and "back stage". Inf Sist E-Bus 7:407–427

Grönroos C, Voima P (2013) Critical service logic: making sense of value creation and co-creation. J Acad Mark Sci 41(2):133–150

Grove S, Fisk R (1992) The service experience as a theater. In: Sherry J, Stemhal B (eds) Advances in consumer research. Association for Consumer Research, pp 455–461

Gustafsson A, Johnson MD (2003) Competing in a service economy: how to create a competitive advantage through service development and innovation. Jossey-Bass, San Francisco

Hoffman KD, Bateson JEG (2010) Service marketing. Concepts, strategies and cases. Cengage Learning, Melbourne

Holmlid S, Evenson S (2008) Bringing service design to service sciences, management and engineering. In: Hefley B, Murphy W (eds) Service science, management and engineering education for the 21st century. Springer, US, pp 341–345

Johnson S, Menor L, Roth A, Chase R (2000) A critical evaluation of the new service development process: Integrating service innovation and service design. In: Fitzsimmons J, Fitzsimmons M (eds) New service development: creating memorable experiences. Sage Publications, Thousand Oaks, pp 1–32

Kimbell L (2009) The turn to service design. In: Julier G, Moor L (eds) Design and creativity: policy, management and practice. Berg, Oxford, pp 157–173

Kimbell L (2011) Rethinking design thinking: part I. Design and Culture 3(3):285–306

Lee E (2016) Service design challenge: transitioning from concept to implementation. In: Proceedings of ServDes 2016, Aalborg University, Copenhagen, 24–26 May 2016

Lovelock C, Wirtz J (2011) Services marketing: people, technology, strategy, 7th edn. Prentice Hall, Upper Saddle River

Løvlie L, Downs C, Reason B (2008) Bottom-line experiences: measuring the value of design in service. Des Manage Rev 19(1):73–79

Mager B (2008) Service design. In: Erlhoff M, Marshall T (eds) Design dictionary. Birkhäuser, Basel, pp 354–357

Mager B, Sung TJ (2011) Special issue editorial: designing for services. Int J Des 5(2):1–3

Manschot M, Sleeswijk Visser F (2011) Experience-value: a framework for determining values in service design approaches. In: Proceedings of IASDR 2011, Delft, 31 Oct–4 Nov 2011

Manzini E (1993) Il design dei servizi. La progettazione del prodotto-servizio. Des Manage 4:7–12

Meroni A, Sangiorgi D (2011) Design for services. Gower Publishing, Surrey

Miettinen S (2009) Designing services with innovative methods. In: Miettinen S, Koivisto M (eds) Designing services with innovative methods. University of Art and Design, Helsinki, pp 10–25

Morelli N (2002) Designing product/service systems: a methodological exploration. Des Issues 18 (3):3–17

Moritz S (2005) Service design. Practical access to an evolving field. http://stefan-moritz.com/welcome/Service_Design_files/Practical%20Access%20to%20Service%20Design.pdf. Accessed 2 Feb 2016

Mulgan G (2014) Innovation in the public sector: how can public organisations better create, improve and adapt? Available via Nesta. http://www.nesta.org.uk/sites/default/files/innovation_in_the_public_sector_how_can_public_organisations_better_create_improve_and_adapt.pdf. Accessed 24 Mar 2016

Nesta (2011) Prototyping public services: an introduction to using prototyping in the development of public services. Available via Nesta. https://www.nesta.org.uk/sites/default/files/prototyping_public_services.pdf. Accessed 13 Apr 2016

Nisula J-V (2012). Searching for definitions for service design—what do we mean with service design? In: Proceedings of ServDes 2012, Espoo, 8–10 Feb 2012

Passera S, Kärkkäinen H, Maila R (2012) When, how, why prototyping? A practical framework for service development. In: Proceedings of the XXIII ISPIM conference, Barcelona, 17–20 June 2012

Polaine A, Løvlie L, Reason B (2013) Service design: from insight to implementation. Rosenfeld Media, New York

Rae J (2007) Seek the magic with service prototypes. Available via Bloomberg. https://www.bloomberg.com/news/articles/2007-09-12/seek-the-magic-with-service-prototypesbusinessweek-business-news-stock-market-and-financial-advice. Accessed 10 May 2016

Sangiorgi D (2004) Il design dei servizi come design dei sistemi di attività. La teoria dell'attività applicata alla progettazione dei servizi. Unpublished Dissertation, Politecnico di Milano

Sangiorgi D (2009) Building up a framework for service design research. In: Proceedings of the 8th European academy of design conference, The Robert Gordon University, Aberdeen, 1–3 April

Sangiorgi D, Maffei S (2006) From communication design to activity design. In: Frascara J (ed) Designing effective communications: creating contexts for clarity and meaning. Allworth Press, New York, pp 83–100

Sangiorgi D, Prendiville A, Jung J, Yu E (2015) Design for service innovation & development final report. Available via Imagination Lancaster. http://imagination.lancs.ac.uk/sites/default/files/outcome_downloads/desid_report_2015_web.pdf. Accessed 5 Apr 2016

Segelström F (2009) Communicating through visualizations: service designers on visualizing user research. In: Proceedings of ServDes 2009, Oslo, 24–26 Nov 2009

Shostack GL (1984) Designing services that deliver. Harvard Bus Rev 62(1):133–139

Steen M, Manschot M, De Koning N (2011) Benefits of co-design in service design projects. Int J Des 5(2):53–60

Stickdorn M, Schneider J (2010) This is service design thinking. BIS Publishers, Amsterdam

Stigliani I, Tether BS (2011) Service design: the building and legitimation of a new category. Working paper—submitted to the Academy of Management conference

Tether B (2008) Service design: time to bring in the professionals? In: Kimbell L, Siedel VP (eds) Designing for services—multidisciplinary perspectives. University of Oxford, Oxford, pp 7–8

Van Oosterom A (2009) Who do we think we are? In: Miettinen S, Koivisto M (eds) Designing services with innovative methods. University of Art and Design, Helsinki, pp 162–179

Weick K (2001) Making sense of the organization. Blackwell, Cambridge

Wetter-Edman K (2011) Service design: a conceptualization of an emerging practice. University of Gothenburg, Gothenburg

Yu E, Sangiorgi D (2014). Service design as an approach to new service development: reflections and futures studies. In: Proceedings of ServDes 2014, Imagination Lancaster, Lancaster, 9–11 Apr 2014

Chapter 4
Exploring Evaluation in Service Design Practice

Abstract So far, the topic of evaluation has been barely treated in service design theory and practice, despite the need to evaluate the contribution of design to innovation which is receiving increasing attention. Starting from the idea of integrating evaluation in the service design process, in this chapter, some examples are provided to show how evaluation could actually foster the design or redesign of better services, and contribute to the discourse on how to determine service design value. Examples are also supported by the opinions of expert practitioners operating in renowned design agencies. What emerges is that evaluation can play a major role in several steps of the service design process, also when undertaken unconsciously by designers, acquiring a different purpose according to the nature of the project and to its application to concepts, prototypes, or full services. Moreover, tools typically adopted in service design practice can be used for evaluation purposes, adapting to the collection and interpretation of qualitative data, as well as to the visualization of evaluation results. This enables a reflection about the possibility to rethink the service design process, integrating an explicit evaluation component aimed at determining the value of designed solutions on the one hand, and guiding and justifying design choices on the other hand.

Keywords Service design practice · Service design approach · Service design process · Service evaluation · Evaluation expertise · Service evaluation process

4.1 From the Evaluation of Existing Services to the Evaluation of Service Prototypes

As said in the previous chapter, evaluation is still an underexplored topic in service design, which is often undertaken unconsciously (as well as service design itself), and it is very difficult to gather meaningful examples illustrating trends, methods, and practices of evaluation in this field. For this reason, we decided to include in this chapter case studies that explicitly adopted a service design approach, and

© The Author(s) 2018
F. Foglieni et al., *Designing Better Services*, PoliMI SpringerBriefs,
https://doi.org/10.1007/978-3-319-63179-0_4

involved some evaluation activities, even though sometimes in an implicit way. Case studies analyzed include:

- Two projects developed by international service design agencies under commission of private firms;
- Two projects developed by institutions operating in the public sector.

The analysis is based on one-hour semi-structured interviews with service designers or project leaders in charge of service design and evaluation activities, and on the analysis of materials they made available to the authors. Interviews included questions about:

- The context that led to the development of the project;
- How service design has been applied and which tools have been adopted;
- Why and how evaluation has been conducted;
- Which are the competencies involved;
- What is the impact of evaluation on the design process and the results achieved.

All interviews were recorded and transcribed, and contents presented as follows have been verified with interviewees. Thanks to these examples it is possible to start a reflection on the role of evaluation in the service design process, with respect to phases in which it could be undertaken and for which purpose, and the approach to evaluation adopted by service design practitioners. What is interesting to notice is that in every case evaluation acquires a different meaning, and it is applied at various moments of the service design process, also adopting different approaches and tools.

4.1.1 'BancoSmart': A User-Friendly, People-Centered ATM

Data and information come from an interview with Jan-Christoph Zoels (Creative Director and Senior Partner of Experientia), and from documents he shared with the authors.

'BancoSmart' is a highly legible, easy to use ATM provided by UniCredit bank that offers a renewed experience of ATM services and personalized options to customers through a responsive design solution developed by Experientia. Experientia is an international experience design consultancy based in Turin, Italy. UniCredit S.p.A is one of the largest bank players in Italy and Europe, with more than 40 million clients in over 22 countries.

Background. 'BancoSmart' is the result of a redesign project of the ATM experience of UniCredit bank. The project started in 2011 with the aim of increasing the use of ATM services for the Italian customer segments, maximizing service findability and usability. The use of this kind of service is low in Italy if compared to the European context, causing an overload of activities for bank branches. The purpose of the project was thus to redesign the information architecture and the navigation of ATM services, based on the analysis of needs and expectations of different final users. The overall project lasted for about 2.5 years, from 2011 to 2013. The new ATM was implemented in 2013 and was rolled out progressively, starting with 6600 devices and covering all Italian branches by the end of 2014.

The design approach. Starting from the brief given by the client and successively refined by Experientia, the project was developed through different steps entailing user research, concept development, prototyping, and user-acceptance testing, up to the implementation of the final solution.

As the first step of the process, designers carried out in-depth user-experience research, using several ethnographic techniques (such as stakeholders interviews, shadowing and talk-aloud observations of current ATM users). They also conducted heuristic evaluation of existing UniCredit and competitor interfaces and organized design workshops to share with the client research insights. The research phase was crucial to gather meaningful qualitative data to redesign the information architecture and to align it with customers' mental models. Moreover, interviews with client stakeholders and workshops supported an in-depth understanding of expected users' interaction with the ATM technology. This phase allowed understanding the barriers in using ATMs and describing users' behaviors when interacting with the machine in different contexts, including for example wayfinding mechanisms of information and functions, or people's perceptions of the interface. User observations allowed designers to collect different data to demonstrate that the low use of ATMs not only depended on the interface design but also on a system of elements regarding a wider idea of interaction. These results fostered more strategic UX decisions during the following concept development phases. Card-sorting sessions were conducted to redesign the information architecture, and initial concepts were tested and refined through physical and digital low-fi prototypes, which also involved final users.

When the final concept was defined, ATM prototypes were further tested in branches using, at the beginning of the experimentation, existing devices. After multiple cycles of design, prototyping and user-acceptance testing, the new interface was finalized so to have a responsive design for different users. The final result can be considered a collaborative solution developed by Experientia and UniCredit and verified through structured meetings during which the ideas were discussed and refined throughout the entire process.

Competencies involved. The Experientia design team in charge of the project was multidisciplinary. At the beginning of the process, it was composed of four people: a project manager, a designer, and two design researchers. During the concept development phase, the team was enriched with other competencies, necessary to the elaboration of the final solution. These included information architects, service designers, interaction designers, and usability testers. From the client side, the process involved those responsible for sales and banking, and three analysts. These competencies were also flanked by legal experts and brand managers to examine all the technical issues concerning the implementation phase.

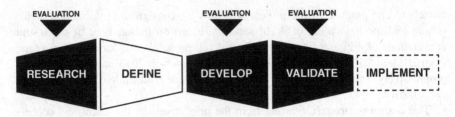

Fig. 4.1 The role of evaluation in the 'BancoSmart' project

The role of evaluation. In this project, the purpose of evaluation was twofold: on one hand, at the beginning of the design process, it aimed at building evidence about service findability and usability of the new idea of a smart and user-friendly ATM; on the other hand, through prototype testing, it supported managers and designers in providing the most efficient and reliable interface to be launched in the market.

The evaluation was introduced at different stages of the design process (Fig. 4.1): first of all, a quick heuristic evaluation with informal participants allowed refining the brief. This was followed by a more structured evaluation dedicated to highlight and assess service values connected to ATM features, according to emerging users' needs and expectations. During concept development, evaluation focused on prototype testing (at different levels of fidelity). In the end, customer satisfaction was measured together with the impact of the new solution on the overall service system.

At the beginning of the design process, evaluation focused on individuating some design priorities connected to findability and usability values that the new service needed to embed: these were the meaningfulness, the legibility, and the accessibility of the information during user interaction and experience. These values guided the preliminary research phase, which could arguably be defined as evaluation research, to understand the gap between the existing solution and the idea of a well-designed and more efficient interface. Current solutions (provided by the client, but also by competitors) were thus analyzed, through shadowing and talk-aloud observations, to highlight where the service experience was coherent to findability and usability values, and where, on the contrary, incoherence emerged. This allowed identifying elements to be improved in order to achieve an easy-to-use interface, a faster understanding of contents, and simplified access to services that take into consideration a diversified community of users. For this reason, this phase involved different kinds of users (such as e.g. elderly people) in different cities and focused on observing users' behaviors to reveal critical issues concerning the navigation, the graphic interface and its look-and-feel, and the ATM functions. The results of this phase were used by the designers to structure the solution and build the first prototypes.

Conducting evaluation at the very beginning of the design process was crucial to foster more strategic decisions. For example, it allowed proving to the client that a geolocalization feature would have increased findability of local/regional services, and facilitated service uptake and upsell; or that the use of company-suggested colors for the interface generated low readability of messages in capital letters. Pieces of evidence produced by evaluation research gave substantial support to designers' proposals, also justifying in some cases the necessity of an extra investment from the client.

Afterward, evaluation was conducted during concept development, in order to test prototypes. Other usability tests were carried out to gather qualitative feedbacks regarding the users' learning process in navigating the new interface, the readability of messages, the ability to read the text, and the effectiveness of the new ATM layout. During *beta testing*, a dedicated team carefully observed and interviewed users interacting with the new device. After multiple cycles of prototyping and user-acceptance testing a new release was prepared and validated through customer satisfaction surveys. These involved more than 60 users showing high satisfaction levels with withdrawal functions, findability of previously hidden services and a perceived decreased time for everyday tasks completion. At this stage of the design process, evaluation results allowed designers to further improve look-and-feel elements such as the final graphic interface, usability issues, and to add new functions and features enlarging the offering of the ATM service without invalidating the clearness of the message and the speed of use.

Final results of design and evaluation. Important improvements have been added to the original ATM interface: some are related to technological aspects (e.g. transition to full-touch interaction), others concern the navigation (e.g. quick access to common tasks, better menu organization), and others are related to contents (e.g. personalization of withdrawal amounts, geolocalized service offerings and advertising). The final solution appears intuitive and easy to navigate, and it also includes new functions:

- A speedy withdrawal with three predefined withdrawal options on the ATM home page. The system now learns over time the most frequent withdrawal amounts of the users and automatically changes the options to be shown. This entails cutting the time for tasks completion by up to 30%, and personalizing the main menu offer to the clients fosters a sense of engagement with the bank.
- A georeferenced payment service organizes bill payment options by filtering them on what is actually available, and showing the relevant information immediately to the users;
- An adaptive interface offers personalized content based on the user's banking profile. This allows providing tailored advertising without interfering with the main activities. The client can touch the right-hand column to access more information;
- A precise system of feedback keeps people informed during the interaction, and helps them to navigate the interface in case of errors.

A multiple transaction offering is introduced that does not require reinserting the bank card. Also in this case, the system offers different options to make the interaction more customer oriented. Thanks to the introduction of new functions and the increased facility and speed of use, new ATMs encourage users in facing other services, offered by the machine, that go beyond basic withdrawal functions. Reducing the time of transactions up to 30% has also generated a positive impact on Return on Investment (ROI). Moreover, UniCredit has increased the usage of their ATMs by more than 25% by non-UniCredit customers, establishing itself as a first-mover in innovative ATM technology.

'BancoSmart' was selected for the ADI Design Index 2014[1] confirming the importance of adopting a design approach in the innovation journey. This recognition reinforced the role of Experientia as a major player in the Italian design ecosystem and helped UniCredit to be recognized as an innovator in the digital service area. Based on the success of this project, they continued to collaborate to design further personal finance services and strategies based on user-centeredness.

4.1.2   'Test Tube Trip': Improving Patient Safety by Reducing the Number of Sampling Errors

Data and information come from an interview with Per Hanning (Creative Director of Experio Lab and responsible of the project) and from documents he shared with the authors.

'Test Tube Trip' is a project developed by the Diagnostics Division of the County Council of Värmland in collaboration with Experio Lab, a Swedish center for patient-focused service innovation. Experio Lab is part of the County Council of Värmland (Sweden) and it aims at introducing and experimenting design approaches in the healthcare system. The mission is to create an environment where care and design can meet, by using the second as a constantly evolving medium to understand people's needs, build courage within the public institutions, envision and implement new solutions, and facilitate cooperation.

Background. Costs deriving from errors committed by the healthcare staff during the preanalytical phase of blood, tissue and cell sampling, count from about 0.6 to 1.2% of the total healthcare expenditure in Sweden. Accordingly, the County Council of Värmland drops approximately five million Euros per year on such problems. Errors connected to blood and tissue sampling causes not only delays in performances and an unnecessary waste of resources but also unpleasant or dangerous situations for patients. In 2014 in Karlstad, 7.5% of all the tissue and cell samples analyzed were incorrect: in some cases, for example, errors merely consisted in misplacing the label on the test tube; in other cases, errors were more severe, such as writing the wrong patient number on the sample label. In general, these inconveniences require patients to repeat the same exam twice, attending very long waiting lists, and, in the worst cases, facing misdiagnosis. For this reason, cost

---

[1]ADI is the Italian Association for Industrial Design that annually collects and awards the best products and services of Italian Design. http://www.adi-design.org/adi-design-index.html.

reduction by reducing the number of operational errors becomes crucial, together with the idea of providing services that are more centered on patient safety. In this context, the head of the Diagnostics Division decided to tackle the problem in a structured manner, involving Experio Lab in creating new models for sampling based on these values.

The design approach. The 'Test Tube Trip' project started in 2015 and was structured through a process made of five steps that include preparation, analysis, and design phases (prepare, explore, understand, improve and implement). Following this structure, the 'Test Tube Trip' project began with a preparation phase. It was, firstly, necessary to convince the project owners about the service design method and set the goal of the project. The aim consisted of creating, testing and implementing at least four prototypes for the sampling of blood, cell or tissue to increase patient safety, reduce costs and create a dialogue between the project owners and the diagnostic lab. The preparation phase also included the selection of the place where to do the design research, the engagement of healthcare personnel, the definition of the time plan and the organization of workshops described below.

Based on this, the exploration phase was conducted through some contextual interviews and field observations aimed at understanding the overall activities of people involved in sampling procedures (in particular personnel from the emergency room, the dermatology and gynecology outpatient services, the blood transfusion unit, the urology and obstetrics wards). Thanks to these activities, six *test tube journeys* (three concerning blood sampling, and three concerning tissue and cell sampling) were conducted and then visualized, showing the interactions between test tubes and people involved from the first touch with the patient to the diagnostic lab. All the research activities were documented through research diaries and pictures to allow people easily using information gathered during the next steps of the project.

Based on the results of this first phase, two workshops were organized with the same practitioners involved in the realization of the *test tube journeys*, and also including employees from the Clinical Education Center in Värmland County Council. The first workshop aimed at discussing and mapping insights from research, and identifying possible changes and improvements to be applied to the sampling process. The second one, aimed at co-designing possible solutions to the problems observed and discussed beforehand, was able to reduce the number of preanalytical errors and the waste of resources during sampling operations. Ideas were clustered and voted by participants to select the most promising ones to be implemented in the future. This activity led to the selection and realization of three prototypes. Each prototype was tested in the field. The first prototype consisted of the creation of a group of 'super-users' composed by nurses and doctors aimed at building a common knowledge on sampling. From the research phase a lack of knowledge sharing was detected, to be attributed to the fact that people were not used to talking about their behaviors and practices. The prototype focused on tailor-made training about how to collect venous samplings. The program was tested by all the teams working in the emergency room at the Central Hospital in Karlstad. The second prototype consisted of the creation of a checklist, describing the activities that can be done for each device used for cell and tissue sampling. The idea was to create a trusted environment in which doctors, nurses and other healthcare personnel involved are acutely aware of the overall process to create a safer sampling journey. Each checklist was created during a workshop where participants were asked to share bad practices they faced during the process and figure out how to accomplish them in a safer way. Moreover, each workshop involved a *sampling relay* process aimed at knowledge transfer: each team was asked to invite a person from another team to pass the checklist and the work was done, so to quickly spread the knowledge obtained by the County Council. The third prototype aimed at reinforcing the message about patient safety. A movie was created comparing the safety protocols in hospitals and in airports, where a pilot and a nurse comment and examine the safety procedures and environments from different perspectives. This was a communicative experiment focused on strengthening people's perception about safety issues comparing two experiences in which this topic is crucial, for both providers and users. In addition to the prototypes described above, a further prototype was realized concerning a physical medical device. It is a new product called 'Helge', which can reveal haemolysis in real time, reducing the time for receiving results from the lab, and thus the necessity to repeat the test.

The 'Test Tube Trip' project is concluded, but the County Council, with the support of Experio Lab, is using the prototypes to conduct other activities aimed at spreading the knowledge acquired and introduce safer sampling procedures across the entire organization. In particular, the first prototype has been adopted by the Clinical Education Center and the sampling relay has been implemented in 33 units out of 193 sample-taking units. At the beginning of each workshop, the movie is shown to participants to kick-start the process.

Competencies involved. The project has been developed as a co-design process, where (service design) experts from Experio Lab and medical practitioners from the County Council work together to respond to a specific brief. Experio Lab has set up the service design and evaluation methodology, organized the activities and facilitated the workshop sessions. Each *test tube journey* involved from three to five people: samplers from the wards and practitioners from the diagnostic lab plus one observer from Experio Lab. The same people have also been involved in the workshops. In general, the hospital employees participate in the process bringing their own expertise and experiences in tissue and blood sampling. The co-design approach generated in all the participants a common sense of responsibility, considering—since the beginning of the process—the achievements obtained as a common outcome.

The role of evaluation. Although evaluation was not explicitly mentioned as part of this project by Experio Lab, we can affirm that it was implicitly conducted both at the very beginning of the process, during the exploration phase, and in the end, to validate prototypes (Fig. 4.2). As a matter of fact, all these activities were not aimed at gathering information per se but were addressed to the measurement and achievement of specific values, cost (and error) reduction and patient safety, as evaluation is supposed to do. Thus it can be said that, in this case, the evaluation of

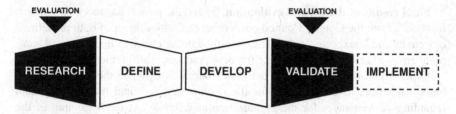

Fig. 4.2 The role of evaluation in the 'Test Tube Trip' project

the existing service enabled the redesign of the service itself, while the assessment of prototypes enabled the refinement of solutions prior to their implementation. Accordingly, specific evaluation (and consequently design) objectives included:

- The achievement of a better comprehension of the different activities that people have to face with when working with samples with respect to the values at the center of the analysis;
- The implementation of solutions that can be accepted and adopted by all the people involved without creating frustration or having an unmanageable impact at organization level (regarding costs and behaviors);
- The reinforcement of knowledge sharing and dialogue between people partici- pating in sampling activities and those involved in the diagnostic laboratories.

The existing service was evaluated from the samplers' perspective through the 'Test Tube Trip' method: personnel involved observed and interviewed colleagues in their contexts and followed the whole journey of the tube, visualizing it through diaries and videos. These materials were shared with the entire community of the project to gather feedbacks and individuate moments and situations when errors happened. This activity allowed people to reflect upon their job and increase awareness of daily practices. What is interesting from the Experio Lab approach is that ethnographic research tools typically adopted in the service design practice were also used here as evaluation tools, which both allowed understanding the weak points of the service concerning the value objectives and fostered a learning process in people involved. As stated above, feedbacks collected were then used as starting points for the workshops, during which doctors and nurses were asked to analyze in depth the *test tube journeys*, reflect on problems emerged and to propose minor changes or new solutions.

The consecutive evaluation of prototypes was still in progress at the time of this interview. The purpose is not only to incrementally change technical aspects of the current procedures, but also the behaviors of people working with samplings. From a quantitative point of view, the pathology department is monthly monitoring the number of errors occurring in cell and tissue sampling and the related frequency of severe errors since the beginning of the project. From a qualitative point of view, Experio Lab is observing behavioral and perception changes toward patient safety in personnel involved in the workshop activities.

Final results of design and evaluation. So far, the project has received positive feedbacks from the County Council concerning the achievement of both predefined objectives and unexpected outcomes, even though the effective decreasing of sampling errors after the introduction of the new practices is still under measurement.

Experio Lab has been able to define a strategy to evaluate the overall service based on values to be achieved for the doctors and the final users, rather than regarding performances for the organization as a whole. At the beginning of the project personnel involved appeared reluctant to changes, since healthcare employees are often trained to follow some routines over time, which lead to the belief that practices are correct just because they have always been conducted in a specific way. Adopting a common language and sharing evaluation conclusions step by step allowed people to create trust in the process and stimulated them to innovate. Moreover, the use of prototypes helped people in visualizing and testing their ideas in the real world, and measuring the useful results. Although prototyping entailed a substantial investment in terms of resources (people, time, money), it was fundamental to introduce a co-design approach in the organization and a self-reflective process by the team involved.

The integration of evaluation elements in the service design processes facilitated participation and collaboration in people involved, enabling the spread of knowledge produced, and awareness about limits and opportunities for the organization. Concerning results, one week after the implementation of the first prototype (training program), the team measured that some preanalytical errors in blood sampling have decreased by 75% compared to the week before.

4.1.3 Designing a 'New Premium Mobility Service' for the US Market

Data and information come from an interview with Stefano Bianchini, strategist and senior service designer at Continuum Milan. The client is kept confidential.

This project was run by Continuum, with a team composed of Milan and Boston designers for a large European automobile manufacturer. The program started in the third quarter of 2012, and the service was launched in the second quarter of 2015. Continuum is a global innovation design firm founded in 1983 working on innovation in products, services, experiences and business models. It has offices in Boston, Milan, Seoul and Shanghai.

Background. A significant change is underway in urban mobility as a result of new consumption models, climate-change issues, and new urban arrangements regarding cars' movement in city centers. Besides, more sustainable mobility solutions are spreading all around the world. Within this new paradigm, the sharing economy is having a disruptive impact on the way people move in urban contexts. Nowadays, car and bike sharing are consolidated practices in many cities, and new players are taking an active role in promoting alternative mobility services and new businesses. For this reason, in 2012, the client contacted Continuum to develop a new premium car rental service to be rolled out in the American market, expanding their traditional role as car manufacturer to one of a service provider. Like most automakers, the client's long-term goal was to get people who don't own vehicles into cars, even if only for a few days.

The design approach. The service design process started with a precise concept coming from the automobile manufacturer about the necessity for the brand to develop a premium door-to-door car rental service. The client commissioned to Continuum the task to structure and design the service features to be implemented. The service design process was structured in three macro-phases that lasted two years:

- Service concept definition (six months);
- Service deployment (nine months); and
- Live testing and refinement (nine months).

The first phase was dedicated to better define the concept, detailing the service value, the core principles and the organization model necessary to deliver the service. An initial customer journey map of the service was outlined to visualize the experience of potential users, which were then described through storyboards. These stories were tested in the field in San Francisco with 15 people in order to receive feedback to refine the service. Results were then visualized through an experience map describing strengths and weaknesses of the service concept.

The second phase mainly entailed the solution development; the design of the digital platform and its components; the definition of the service model necessary to manage, improve and maintain the service; and the support of the role of the product owners and SCRUM[2] team (run by another company) developing the digital touchpoints. During this phase, the team considered different design directions to prefigure solutions for both success and failure situations. They tested design directions in the field, through experience prototyping, involving users in a real-life situation where they were asked to use the service (and real cars) for an entire weekend. At the conclusion of the test, people were interviewed in order to share their experiences. Continuum identified delight factors (namely all the features that people recognized as key success elements), such as the simplicity of the software, the accessibility to the entire fleet of cars manufactured by the client, the presence of an 'agent of service' to communicate with, and so on. The output of the second phase consisted of the finalization of the service blueprint and the service policies, detailing all the interactions with physical, digital and human touchpoints necessary to the management, delivery and use of the service. The service was then effectively deployed in all its components, from software development to the involvement and training of human resources.

The third phase focused on the fine-tuning of the client experience, which was conducted through three rounds of testing, which culminated in a live performance

---

[2]SCRUM is an iterative and incremental agile software development framework for managing product development.

of the service. Continuum designed a feedback tool which was utilized after the end of each round of testing, collecting criticalities both from the staff and the user point of view to understand if emerging problems were related to the process, system, or brand perception, and fix them.

Competencies involved. Different design competencies were involved in managing the different steps of the project, and representatives from the client company participated in field research and other service design activities, thus going beyond their traditional roles. The Continuum team included a service designer, a digital designer, a design strategist, a researcher, and a visual designer (envisioner). The client group consisted of engineers and managers from their Business Innovation Unit coordinated by a project manager from the company headquarters. The evaluation process was coordinated by the Continuum design team as well with the support of a recruiting agency in charge of user engagement for the prototype testing. Software solutions were realized by an external innovation agency, which was also in charge of the SCRUM process and the design of the digital interfaces.

The role of evaluation. In this case, the evaluation process ran in parallel to the service design process (Fig. 4.3), since the design team in charge of the project constantly checked, tested and verified the service solution proposed, running several rounds of testing for different purposes according to the level of development of the solution. In general terms, it can be said that the evaluation aimed to monitor the quality of the service solution from an experiential point of view and to verify the coherence between the service and brand values, namely the reliability of the technology and efficacy.

The evaluation process was thus strictly connected to the different prototyping phases that were structured during the service development, supporting the iterations necessary to modify and validate all the service components, to launch a new service with the least number of problems or bugs. In fact, due to the client's brand reputation, it was not possible to test the service directly on the market through *beta-versions*. In particular, three main evaluation cycles were conducted:

1. The first one was dedicated to validating the service concept, evaluating the level of relevance of the service for the users;
2. The second one was focused on the evaluation of the service experience and the early proposal of the digital interface, concerning user needs and the compliance to brand values; and

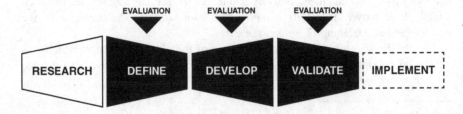

Fig. 4.3 The role of evaluation in the 'New Premium Mobility Service' project

3. The third one evaluated a high-fidelity prototype of the service, which at this stage was fully developed and ready for market launch.

Final results of design and evaluation. The first full premium door-to-door car rental service was launched in San Francisco in 2015. It allows customers to match specific vehicles to their driving needs thanks to a wide range of cars made available by the service. The key feature of this service is that people can decide to pick up the selected vehicle in a particular place or to have it delivered to a location within the city. The service relies on personal smart devices, which facilitate users in booking, accessing, using and returning each vehicle.

Testing highlighted the importance of a human presence to enhance the perception of the brand as premium; therefore, the service includes a particular element: the *agent of service*. During the first use of the service, the agent supports the user in accessing the vehicle, reducing the *feeling lost* perception typical at the beginning of a car rental. Another characteristic developed thanks to the evaluation's results is related to the moment of car return, which has been simplified to the extreme. Consumers leave the vehicle in front of their house (or another location of their choice) and it magically disappears thanks to a precise pick-up mechanism managed by the service provider. After the first period of silent launch, the service is now public, and it's continuously developing: the client has just rolled it out to another location in Europe.

4.1.4 Evaluating the Impact of Service Design Introduction in Mind Mental Healthcare System

Data and information come from an interview with Paola Pierri (Local Services Strategy and Development Manager at Mind) and from internal documents she shared with the authors.

'Embedding excellent service design' was a large-scale research-based project run by Mind in the UK in 2014, which aimed at introducing the service design approach and tools at different levels of the Mind organization. It mainly focused on reinforcing user engagement, providing training on service design and supporting the development of service design activities for this purpose across the Local Minds' network.

Background. Mind is a charity focused on mental health issues based in England and Wales. It is made up of a network of around 140 independent charities (Local Minds). The network serves over 400,000 users affected by mental health problems across the country, providing services such as housing, crisis helplines, counselling, eco-therapy, employment and training schemes, counselling and befriending. Each Local Mind is run by local people for the local community and is also responsible for its own funding. In collaboration with the Innovation Unit, Mind has developed a service design toolkit to be used by Local Minds aimed at enabling the development of mental health services in a truly person-centered way, and in partnership with commissioners, providers, staff, family and friends.

Beyond running initiatives across the Local Minds' network addressed at the introduction of the service design approach, the project also included an evaluation phase. A formative evaluation was firstly conducted during the project to monitor the ongoing achievement of predefined objectives. Another evaluation was then conducted six months after the end of the project to determine the actual benefits and the barriers of adopting service design in this context, participant perceptions in terms of engagement in the service design process, and perceived impacts for participants of working with the service design approach. Considering the purpose of this chapter, we focus here on the second evaluation, which focused on three specific Local Minds. As an example, we describe the initiative held by the Bromley & Lewisham Mind, which has been the object of this evaluation. Bromley & Lewisham Mind participated in the project to improve the service offered by its Wellbeing Centre. The Centre is always open and offers different activities such as peer support, where service users provide their knowledge and emotional support to each other, based on the common lived experience of mental health and with the aim to move together towards recovery. This Local Mind decided to introduce service design activities to investigate users' attitudes and behaviors to better manage the space or individuate possible changes in the current service. In fact, from a previous analysis, they realized that some people used the service in a continuative way for many years without progressing in the recovery process, while others quit the service after a shorter period for other services more suitable to their particular (or improved) condition.

The design approach. The purpose of this project was to integrate the service design approach in the organization to improve the design of new personalized and recovery-focused services around mental health and long-term conditions that are scalable from the local level to the national one. Consequently, the charity developed its own methodology that was co-designed with Mind and the network of Local Minds as well as the support of The Innovation Unit, a collaborative of designers, researchers, public service leaders and practitioners working as an innovation partner for public services. Mind service design process consists of five stages:

- The Set Up phase is about positioning the project strategically, building the team and defining the brief;
- The Explore phase is dedicated to the generation of new insights through ethnographic research;
- The Generate phase concerns the framing of the brief, service ideas generation and early prototyping;
- The Make phase is mainly dedicated to prototyping and testing;
- The Grow phase deals with the implementation of the solutions, including the business plan.

While developing the methodology the co-design team also identified a narrative and an actionable plan useful to communicate the approach to the network and support Local Minds learning process toward the adoption of service design. To exemplify how the process works in practice, we refer to the Bromley & Lewisham Mind experience. The process started with the Set Up phase, during which the brief was defined: redesigning part of the existing offering of the Wellbeing Centre to improve the effectiveness of the service by understanding why some people get dependent from the Centre without progressing in the recovery process. To reach this purpose, they structured a research phase, mainly including interviews done by users of the Centre for other users. Discussions aimed at collecting information about the users' social networks and places they attend to reinforce or develop their social skills. Other research activities dealt with mapping daily routines and behaviors, focusing on people's motivations in attending the Centre. The output of this phase consisted of a series of personas describing different typologies of users and generated a first understanding of strengths and weaknesses of the service design tools adopted. Insights emerged from the interviews were then clustered and organized to describe promising redesign areas. In the end of the research phase results were shared with local funders to find possible strategies to implement ideas identified.

Competencies involved. The project involved different competencies regarding mental health, but also design and evaluation. A coordinator with service design competencies from the Mind office supervised the whole process. The internal Mind service design team was also in charge of training Local Minds on how to conduct user research and analysis and to structure the ethnographic activities. Experts from The Innovation Unit, working together with Local Mind's employees and users, proposed the first draft of the service design methodologies. The Point People, a London-based consultancy working on building communities, mobilizing movements and sustaining networks, ran formative evaluation during the project. The assessment managed after the end of the project was conducted by ResearchAbility, an independent qualitative research agency.

The role of evaluation. As mentioned above, the role of evaluation in this context was to measure the capacity and readiness of Local Minds to embed and use service design and its impact on the organization and service users, understanding the adoption barriers and the related challenges (Fig. 4.4). Focusing on the evaluation conducted by ResearchAbility after six months from the conclusion of the

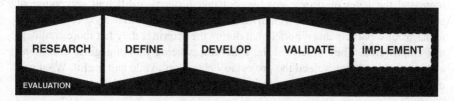

Fig. 4.4 The role of evaluation in Mind's 'Embedding excellent service design' project

project on three Local Minds, three focus groups were conducted with a total of 15 service users who had been involved in some aspects of the service design methodology developed by Mind. General value objectives at the core of the evaluation process were mainly assessed by exploring:

- Users' experience of the service design toolkit, including barriers and enablers to effective use;
- Users' perceptions of engagement in the service design process;
- Perceived impacts of working with the service design approach;
- Recommendations and lessons learned.

Each of the three Local Minds involved convened the focus groups, which were facilitated by ResearchAbility with an agreed topic guide. Panel discussions were recorded to analyze data, which were summarized into a thematic framework.

Despite some meaningful insights being captured during the evaluation process, some limitations must be recognized:

- At the time of evaluation, the three Local Minds involved had only used one or two stages of the service design methodology or some of the service design activities.
- The evaluation only considered the perspective of service users and not the viewpoint of the service provider or the service staff.

Moreover, we could add that focusing on each Local Mind initiative it could have been interesting to observe how the research phase of the service design methodology proposed can support the formulation of evaluative conclusions regarding the effectiveness of existing services and thus addressing the development of new solutions. For example, looking specifically at the initiative undertaken by Bromley & Lewisham Mind, results of the ethnographic research conducted by users on the Wellbeing Centre enabled the generation of improvement ideas, promoting a change in the organizations not only regarding service design capabilities acquired but also in terms of transformative power.

Final results of design and evaluation. Results of the evaluation conducted by ResearchAbility concerns the development of service design capabilities in Local Minds and the impact of the service design methodology, especially in terms of user engagement.

Despite the meaning of service design and the use of service design terminology not being fully captured by participants, it can be said that from a user perspective the participation in service design activities has had a positive impact. It has increased the level of people engagement and fostered a different perception of being part of the organization as individuals with different stories and needs. During focus groups, participants reported that during the activities they felt more confident and empowered. They described the activities as being fun, engaging, energizing and creative, and they valued the project as being worthwhile and useful. What was mostly highlighted was their perception of having contributed to the progress of the projects.

Moreover, concerning the impact of the project on the organization, Mind has identified three main lessons learnt from the overall process:

- Build a network of support, namely structure a community of people able to share competencies with external networks involved in similar processes to avoid frustration in people who do not (yet) fully understand the benefit of the adoption of service design;
- Build the right knowledge about the service design landscape in the mental health field to nurturing positive relationships with other partners and subjects;
- Enable people to adopt service design in Mind's programs in order to show its impact on traditional Key Performance Indicators (KPIs)[3] and engage people in including service design practice not as (just) something new to do but as a new organization's offer.

Since the definition of the Mind service design methodology the team has developed a theory of change and a plan for how they want to embed a design culture in the organization. They have currently actively engaged with 45 Local Minds on different projects, 10 of which will be intensively supported through the 'Grant +' system from April 2017. The 'Grant +' system provides funding to Local Minds to go through a phase of the design process (mainly the phase Explore and Make) but the capacity building element represents its innovative edge. The + of the scheme is, in fact, a training package and a design coach that each Local Mind has got assigned, in addition to the funding. This has proved a very successful way so far to ensure that Local Minds have capacity and resources to run a design project while *learning by doing*, and with the support of an external design coach, who plays the role of the critical friend.

Through product and program development, Mind has also exposed eight more Local Minds to the service design approach. Examples of the service design offer for the Mind network include:

- A full day 'service design sprint' to introduce design techniques by working hands-on on a real project for a day;
- One to one advice and support for specific projects;
- Research training for staff and service users to start familiarizing with design research techniques.

The 'Grant +' scheme has also been replicated with funding and coaching provided to more projects that were going through the research phase or to prototype service ideas.

For what concerns the implementation of the redesign solution at Bromley & Lewisham Mind, the process was still ongoing and information was not available at the time of the interview.

---

[3]Set of measures focusing on those aspects of organizational performance that are the most critical for the current and future success of the organization (Parmenter 2010: 4).

4.2   Evaluation in Service Design Through Professional Lenses

Before reflecting on what emerges from the analysis of case studies, we report the points of view of some experts about the meaning and role of evaluation in service design practice. In particular, we had the chance to interview two professionals with a management role in international design agencies: John Oswald, Business Design Director at Fjord, and Gianluca Brugnoli, Executive Director of User Experience at frog design in Milan. Also in this case, we carried out a one-hour semi-structured interview, but focusing instead on how evaluation is perceived within their organizations, how it is conducted in their professional practices and for which purpose, and how, in their opinion, it affects the service design process.

Fjord, as well as other major design agencies such as IDEO, is developing a robust methodology that links *business design*[4] and *design thinking*. In the opinion of John Oswald applying design principles and skills to business as a problem-solving mindset that helps to face real-world changes is the next big challenge for industries. In such a business-focused context, he considers evaluation contribution to design strictly connected to the definition and measurement of KPIs. Examples are the list of outcomes to be achieved by a project in relation to the overall brand and customer value (e.g. impact on future revenues, create a service that helps the brand repositioning in the mind of customers, etc.). And it is also about identifying values and principles that will guide market analysis, user research, and design ethnography, and that should characterize the service to be developed. According to this perspective, evaluation represents a core competence of the *business design* team at Fjord, which gives more consistency to the design process concerning the overall purpose of the design intervention in terms of customer and company values to be achieved. At the same time, it extends the understanding about how the brand engages with its customers. This point, in particular, represents the added value of including evaluation in service design projects, which goes beyond the measurement through financial metrics typical of business organizations.

In synthesis, as also emerging in some of the cases analyzed before, Oswald highlights two different steps of evaluation along the design process:

- The first one is related to the preliminary definition of values and principles useful to guide the research phase, and measure solutions in term of business success;

---

[4]*Business design* can be considered as an approach to innovation that merges the human-centered approach, experience design and business strategy. It applies *design thinking* to help organizations creating viable solutions and competitive advantage (see e.g. Martin 2009; Slywotzky et al. 2002. *Business design* is not widely explored by literature, but design agencies are increasingly requiring these kinds of competences, contributing to the consolidation of *business design* as an approach handled by a specific professional figure.

- The second one is related to the design phase and entails the definition of filters through which one prioritizes ideas and identifies those more aligned to the brand, delivering customer and business value in short and long term. The same filters should also accompany the service development phase, supporting design decisions step by step.

On the other hand, Brugnoli suggests that there is not a unique way to evaluate services; he affirms that the evaluation strategy to be undertaken depends on the scale of the project, on its objectives, and other contextual factors. Moreover, it is firstly necessary to clarify if the evaluation is related to the final user or to the organization, as well as how to mix quantitative and qualitative methods and data. It is also the opinion of Brugnoli that when the evaluation is addressed at the organization, KPIs play a predominant role. Among these, he mentions, for example, the conversion rate as one of the most diffused, connected to the so-called *moments of truth*.[5] However, from his perspective, in service design practice evaluation is mainly related to qualitative analysis through ethnographic methods (e.g. interviews, shadowing, diaries, etc.) aimed at understanding people's needs and behaviors. Quantitative analyses are usually transferred to marketing competencies.

Referring to the service design process, according to Brugnoli, evaluation can be applied in several steps, depending on the nature of the project. In general, at an early stage of the process, evaluation allows understanding how the service works and how people interact with it. While during the design phase, evaluation supports the validation of ideas, through prototyping and user testing of the final solution. Different tools can be adopted for these purposes, although he considers the *customer journey map* the most powerful tool for qualitative analysis, also when aimed at evaluation. In fact, mapping the customer journey at the very beginning of the process can help designers and managers to frame the current situation of the service, highlighting the potential areas of intervention as well as critical points. Moreover, during idea generation, it can support the prioritization of problems and help figure out alternative solutions. In the end, the *customer journey map* can also be used to detail solutions, going deep into organizational procedures and protocols, user interactions, and the design of related touchpoints.

4.3 Toward the Integration of Evaluation Expertise

Case studies described in this chapter represent different kinds of service design projects as well as various ways of applying service design competencies within organizations. The 'BancoSmart' project and the 'Test Tube Trip' show how service design can support the improvement of existing services: in the first case

---

[5]The term moment of truth has been introduced by McKinsey & Company and is usually defined as the instance of interaction between the customer and the organization when the customer forms or changes an impression about the firm.

guiding the redesign of a key touchpoint for the bank branches, and the related service experience; in the second case favoring the optimization of processes through the development of solutions aimed at educating and increasing collaboration among healthcare personnel. The third case consists instead of the design of a brand-new service aimed at expanding the business of a manufacturing company. This provides an example of how service design can answer to the need for *servitization* (see Sect. 2.2) by putting into play specialized approaches and competencies, but also working as an orchestrator for other kinds of expertise and implementation processes. Lastly, the fourth case introduces another level of discussion about the possibility to embed service design capacity within organizations, in this case as a medium for user engagement and co-design, and the necessity to measure its impact to demonstrate its real contribution to service improvement.

From a service design point of view, regarding what was reported in Chap. 3, in all these cases, service design interventions are characterized by a prearranged process wherein the initial research step played a significant role. It is demonstrated that qualitative analysis based on user needs, expectations, and perceptions constitutes a substantial basis to drive the design of solutions and service improvement. In the 'Test Tube Trip' project, it becomes useful also to investigate the human dynamics behind organizational processes; while in Mind healthcare system users are trained to set up and conduct ethnographic research activities by themselves, reinforcing the value of this phase of the process as a crucial aspect of embedding service design capacity. On the other hand, quantitative analysis acquires a limited role, which is transferred to other professional figures, as suggested by Brugnoli (see Sect. 4.3).

Some similarities can also be detected in the following steps of the service design process. With the exception of the 'New Premium Mobility Service' project that did not include a concept generation phase, in the other cases, generative sessions have been conducted through workshop activities involving decision-makers and staff from the organization, or users as in the Mind example. Another similarity is given by the need of validating service concepts through prototype testing (even though in the 'Test Tube Trip' project prototypes focus on education formats rather than service solutions). Moreover, 'BancoSmart' and the 'New Premium Mobility Service' projects show that the validation phase does not necessarily confine to a single moment of the process (i.e. after service idea development as illustrated in Figs. 4.1 and 4.3), but can be also located in the end of the concept generation phase to support the selection of the best ideas. Further, it can be iterated several times within the development phase, especially when dealing with digital components. Concerning the implementation phase, these examples confirm that, even though it is acknowledged as a final step of the service design process, it is delegated to the organization.

Analyzing these examples, some common aspects and tendencies can then be recognized from an evaluation point of view, which makes us understand how assessment in service design is interpreted nowadays. These concern the positioning of the evaluation in the service design process on one hand, and the levels of its application on the other hand. Looking at Figs. 4.1, 4.2 and 4.3 it clearly emerges that the evaluation can occur at different moments of the service design process:

- At the very beginning, that is during the research phase, to build knowledge about the context in which the project takes place, or the service that needs to be redesigned;
- During service idea development and validation (and eventually also as the conclusive act of the concept generation phase), through the evaluation of service prototypes to select and refine solutions, and verify if they work as they should before being implemented.

Thus, it could be said that in the service design process evaluation coincides with (ethnographic) research and prototype testing, but it is not merely a matter of naming things differently. It is rather a matter of value.

As better explained in the next chapter, evaluation equals determining the value of what is evaluated, either a concept, a prototype, a service or service design itself. As a consequence, talking about evaluation applied to research and prototype testing means shifting the purpose of these activities from producing a factual knowledge (how things work) to an evaluative knowledge (if and how things answer to given values). This clearly emerges from all the examples analyzed (e.g. in the 'BancoSmart' project the purpose of research was to assess the findability and legibility of information in the ATM interface) and it is fundamental to develop solutions that bring value to the user and the organization (in agreement with innovation requirements), and provide evidence about the value of service design interventions. If we consider the case of service redesign, the importance of evaluation becomes even more explicit. In fact, conducting service evaluation at the beginning of the service design process allows building the necessary knowledge with respect to the current value of the service to be redesigned, and basing its transformation on a robust and shared evidence. Then, evaluating the redesign solution implemented allows measuring the improvement generated in comparison with the starting situation and related value objectives.

The positioning of evaluation at these steps of the service design process, and the importance of identifying value objectives to be achieved by the project, is confirmed and emphasized by the interviews with experts (see Sect. 4.2). They both relate evaluation to the definition and measurement of KPIs (that we also find in the Mind project) as the most common way for organizations to set their goals in terms of value. It is also interesting to notice that Oswald refers to evaluation as an activity that can give consistency to the design process, in agreement with our idea of integrating this expertise in service design practice in order to build evidence on solutions developed.

Another thing can be observed about the application of evaluation in service design practice. Based on our case studies, three main areas can be identified:

- Evaluation of existing services in order to better address the service design intervention, such as in 'BancoSmart' and 'Test Tube Trip' projects;
- Evaluation of service design solutions, at concept level or at a more advanced level of development, such as in 'BancoSmart' and the 'New Premium Mobility Service' projects;
- Evaluation of service design impact at organization level (Fig. 4.4), such as in the Mind project.

This is in tune with the idea of Brugnoli that there is not a unique way of conducting the evaluation in service design, depending on the scale and the objectives of the project.

In general, we can recognize the tendency to evaluate at the project level, also when the intervention consists in redesigning a service. In fact, in these cases, evaluation focuses on the existing service and the prototype of the solution developed, rather than the existing service and the new or improved service once it has been implemented, taking for granted that the solution will prove successful. The same happens when the purpose of the project is to design a new service, such as in the 'New Premium Mobility Service' project, where evaluation concentrates on the evaluation of prototypes, but the solution is no more evaluated (at least by service designers) once implemented. This can probably be associated to the marginal contribution of service design to service implementation, emphasizing the need to further explore how to approach this step in the service design process. On the other hand, much effort is still being put into demonstrating the value of the service design approach, so as to justify the organization's investments, as in the case of the Mind project.

To conclude, the analysis of case studies shows that service design tools can be used for the collection and interpretation of qualitative data, as well as for the visualization of evaluation results. In particular, the *customer journey map* seems to be a powerful instrument to support evaluation activities too. This opens up another stream of discussion about how to identify and adapt service design tools for evaluation. Those illustrated in this chapter are just some examples regarding the integration of evaluation expertise in service design practice. We must admit that service design proves to be endowed with a transversal capacity that can be applied to broad project requirements, adapting its approach and tools to practices that are related to but do not coincide with design, such as evaluation. On the contrary, it is highlighted that awareness and a conscious vision that recognizes evaluation as a process useful to support the design practice and generate more valuable services still needs to be built.

Based on the capacity of service design to approach evaluation on one hand, and the need for an aware vision on the other hand, in the next chapter we go deep into the topic, firstly exploring the meaning and context of evaluation applied to services, and then proposing operative guidelines to put theory into practice, and start considering evaluation as a *must do* in service design practice.

References

Martin RL (2009) The design of business: why design thinking is the next competitive advantage. Harvard Business Press, Boston

Parmenter D (2010) Key Performance Indicators (KPI), developing, implementing and using KPIs. Wiley, Hoboken

Slywotzky AJ, Morrison DJ, Andelman B (2002) The profit zone: how strategic business design will lead you to tomorrow's profits. Three Rivers Press, New York

Chapter 5
Evaluating Services for a Better Design

Abstract To understand how to integrate evaluation in service design practice it is first fundamental to fully understand the importance of evaluating services in the contemporary context and what it means to evaluate. To do so, we refer in particular to program evaluation literature in the field of social sciences, since these types of studies were the first to approach the conceptualization of evaluation, structuring methodologies to design and conduct an evaluation strategy. Based on this knowledge, the chapter focuses on the evaluation practice in the service sector and the related concept of service value, with the purpose of translating existing expertise into the service design practice. This implies a necessary distinction between the evaluation of services (being considered as both service concepts or solutions not yet implemented and services resulting from service design interventions or object of redesign projects) and the evaluation of service design as an approach that can bring value to organizations. While the first can eventually support the latter, determining the value of service design does not give any evidence about the value of services it delivers in a particular context. Thus, a definition of service evaluation as a process that intersects and substantiates the service design process is formulated and a conceptual model that supports the design of a service evaluation strategy is proposed. This includes key elements to be negotiated by designers and decision-makers and a set of guidelines useful to pragmatically conduct a combined service design and evaluation process.

Keywords Evaluation · Service evaluation · Design evaluation · Service design evaluation · Program evaluation · Evaluation research · Evaluation strategy · Service evaluation process · Service evaluation guidelines

5.1 Why Evaluating Services?

Before going deep into the meaning of evaluation and exploring its role in service design practice, we want to explain further why it is important to evaluate services (from a service design perspective) and how evaluation is conducted in the service sector.

© The Author(s) 2018
F. Foglieni et al., *Designing Better Services*, PoliMI SpringerBriefs,
https://doi.org/10.1007/978-3-319-63179-0_5

Service organizations are increasingly realizing that applying design competencies to the development of services can result in stronger brands, improved customer satisfaction, the acceleration of new ideas to market, or the creation of new markets, enhancing service innovation (see Chap. 2). The same happens in the public sector where (service) design is acknowledged as a new approach to design policy, reshaping professional practice and exploring new disciplinary territories (McNabola et al. 2013). However, we must consider that new challenges are appearing at the service design door. We are experiencing change on a scale and level of complexity never encountered before. The adoption of service design can support the delivery of valuable services that answer and adapt to contemporary challenges. This entails for service design to broaden out to new practices and competencies such as adopting an evidence-based approach (Carr et al. 2011) and measuring the value of services it delivers, in a continuous innovation perspective that is coherent with the complexity of the current socio-economic context.

As shown in Chap. 3, the approach, methods, and tools to be used in designing services have been extensively discussed and developed in the last two decades. However, the lack of rigorous theory and principles are still perceived, and service design continues to be undertaken unconsciously by organizations, often restricted to the use of some tools. This turns out in the overlapping of service design and *design thinking* approaches, or worse, in the creation of services that quickly become uninteresting and obsolete, or that replicate tons of similar services, the majority of which never manage to achieve the critical mass necessary for their economic sustainability. In this context, the evaluation of services acquires a strategic role. It allows providing evidence on what works and what does not work in a service, understanding processes and practices behind service models and, consequently, addressing design interventions that are based on a solid, shared and shareable knowledge, as well as proving the value of services resulting from design interventions.

In our opinion, the understanding of service value, and knowing how to measure it, is what is lacking in the service design practice to face the challenges posed by the service economy and society. Both service providers and designers need to monitor and improve the value of the services they design and deliver. They need to provide reliable and shared evidence upon problems identified and solutions proposed, which in turn must be coherent to the provider's values (and to the brand value), to the user's values, and ideally to the context's values. We argue that this can be done introducing evaluation in the service design practice. Why? The reason resides in the meaning of evaluation itself. The evaluation must not be intended as the expression of a judgment per se but as the expression of a judgment that enables a critical process of learning and change. Evaluating systematically helps people and organizations to dramatically improve their work, supporting decision-making processes (Scriven 1991; Shaw et al. 2006). It builds knowledge and skills, developing a capacity for evaluative thinking that facilitates continuous quality improvement, providing evidence about the value of what is evaluated, and justifying interventions undertaken (Donaldson and Lipsey 2006).

The importance of evaluating programs and policies, as well as products, services, and processes lies in the need of governments and firms to keep on responding to customers and communities' needs and expectations according to the evolution of the welfare state (Gamal 2011; Gillinson et al. 2010; OECD 2011). It answers to the demand for evidence-based decisions, but also to evaluate innovation as an indicator for benchmarking public performances (see e.g. 'Frascati Manual', 'Oslo Manual', 'European Innovation Scoreboard', etc.). Thanks to rigorous evaluations, funders, policymakers, as well as service providers can understand what programs and services accomplish, how much they cost, how they should be operated to be effective, and how they respond to emerging needs.

Not surprisingly, the evaluation field has expanded a lot in recent years. Living in a historical moment characterized by financial difficulties and scarcity of resources both in business and the public sector, it is more and more frequent to hear about evaluation (Forss et al. 2011), especially related to programs and policies at local, regional and national levels. In fact, in 2015, the global evaluation community celebrated the 'International Year of Evaluation' aimed at advocating evaluation and evidence-based decision-making at all levels. Thus, we propose evaluation as a support practice to service design, able to provide the necessary knowledge to address valuable design solutions, and enabling a continuous *learning–design–change cycle* as required by the contemporary context. From a design point of view, spreading a culture of service evaluation could eventually increase providers and users awareness on the importance of service design as a driver of innovation too (Foglieni et al. 2014). However, we must not forget that evaluating is a complicated process that implies responsibilities and a research aptitude, and that evaluation practice in the service sector, and service design, in particular, is still a fragmented and controversial issue.

To build the necessary knowledge to address the topic, it is necessary to explore extra-disciplinary fields of study. We start with the concept of evaluation to then explore how it applies to services, showing how service evaluation and service design can benefit from each other in order to design useful and efficient services that are coherent to the broader context in which we live and to introduce meaningful innovations.

5.1.1 Exploring the Meaning of Evaluation

The most common definition of *evaluation* reads to judge the value of something or someone in a careful and thoughtful way (Scriven 2007). However, without searching for a definition, everybody knows what evaluating means, since throughout our lives we constantly experience evaluation. We are evaluated at school, at work, by people we meet, and we evaluate as well every time we make a choice or express an opinion (Foglieni and Holmlid 2017). In few words, evaluating means expressing a judgement and this is the main reason for it being often a cause of prejudice, because being judged is not always pleasant. However, in this book, what we refer to by the term evaluation is not that kind of informal (or tacit) evaluation we make in our everyday life, but rather systematic formal evaluation conducted by professional evaluators. While informal evaluations do not explain what has led to a particular conclusion, formal evaluations make explicit the evidence and criteria on which judgments are based, using codified methods (Hammersley 2003).

> Systematic evaluation can offer a way to go beyond the evidence available to any individuals, as well as to facilitate evaluative processes that are collective and not simply individuals (Shaw et al. 2006: 2).

Systematic evaluation allows understanding what works (or not) about activities, performances, programs and projects—including services—in order to replicate or refine them (Bezzi 2007). If an activity is positively evaluated it can be reproduced or inspire other initiatives; if it has produced bad results, evaluation makes us understand what went wrong and fix it, avoiding repetition of the same mistake.

Systematic evaluations can be applied to several contexts, situations and activities (Morrison 1993). However, the formalization of systematic evaluation theory and practice were born concerning programs and policies, in the field of social sciences (mostly from education, social psychology, sociology, economics, and political sciences). This is mainly due to historical reasons because the need for transforming into theories the evaluation practice was born as a consequence of the social and political demand for valuing public programs after the 1930s US Great Depression, which forced social sciences into service (Mathison 2005). And to the fact that programs and policies usually involve the evaluation of people, performances, projects, products, services and other aspects, adapting to many fields of application (Shaw et al. 2006). According to program evaluation literature, evaluation can be defined as

> An applied inquiry process for collecting and synthesizing evidence that culminates in conclusions about the state of affairs, value, merit, worth, significance, or quality of a program, product, person, policy, proposal, or plan. Conclusions made in evaluations encompass both an empirical aspect (that something is the case) and a normative aspect (judgment about the value of something). It is the value feature that distinguishes evaluation from other types of inquiry, such as basic science research, clinical epidemiology, investigative journalism, or public polling (Fournier 2005: 140).

From this definition, a crucial aspect emerges: what distinguishes evaluation from other research practices is the value feature. From an etymological point of view, the term *value* refers both to finding a numerical expression for, and estimating the worth of. It can be defined as a perception (Woodruff and Gardial 1996) or a feeling (expressed in a quantitative or qualitative manner) we attribute to something.

Starting from the understanding of these key concepts, it is then important to depict what the activity of evaluating consists of when facing social programs,[1] or better, planned activities as also services can be intended. The definition of *program evaluation* can be useful in this sense:

---

[1]Social programs are defined as organized, planned and usually ongoing efforts designed to ameliorate a social problem or improve social conditions (Rossi et al. 2004).

> Program evaluation [is] a social science activity directed at collecting, analyzing, inter-
> preting and communicating information about the working and effectiveness of social
> programs. [...] Evaluators use social research methods to study, appraise, and help improve
> social programs, including the soundness of the programs' diagnoses of the social problems
> they address, the way the programs are conceptualized and implemented, the outcomes they
> achieve, and their efficiency (Rossi et al. 2004: 2–3).

According to this definition, the activity of evaluating must be intended as a critical
process, based on data collection and interpretation, aimed at triggering a change,
an intervention that allows improving or replicating what is being evaluated. The
purpose of this activity is not to highlight a fault or negligence, nor to stigmatize;
through research it builds a reasoned hypothesis about the value of what is being
evaluated, which goes beyond the idea of controlling the legitimacy (or accountability)[2] of the program at hand (Chelimsky 2006; Feinstein and Beck 2006). It is
rather to be considered as a learning and transformation activity able to foster a
culture of change (Bezzi 2007; Grietens 2008).

Research is the key aspect that distinguishes informal evaluation from systematic
evaluation. Not by chance Scriven (1991) suggests that *evaluation research* is just a
name for serious evaluation, but in fact talking about assessment coincides with
talking about evaluation research. Many scholars refer to social research when
defining (program) evaluation (see e.g. Patton 1997; Palumbo 2001; De Ambrogio
2003), but even though closely related, evaluation is a methodological area that
differs from more traditional social research. Understanding this difference is fundamental also to figure out how ethnographic research in service design practice
may acquire another value when conducted for evaluative purposes. In an interview
for an issue of the *Evaluation Exchange*[3] (Coffman 2003–2004), Scriven addresses
the issue asserting that

> Social science research [...] does not aim for or achieve evaluative conclusions. It is
> restricted to empirical (rather than evaluative) research, and bases its conclusions only on
> factual results—that is, observed, measured, or calculated data. Social science research does
> not establish standards or values and then integrate them with factual results to reach
> evaluative conclusions. In fact, the dominant social science doctrine for many decades prided
> itself on being value free. So for the moment, social science research excludes evaluation.
> However [...] without using social science methods, little evaluation can be done.

Thus, the main difference between evaluation and social research resides in their
purpose. While evaluation requires the synthesis of facts and values in the deter-
mination of merit, worth or value of something, the research investigates factual
knowledge but may not necessarily involve values, thus not including evaluation
(Mathison 2008: 189). In other words, while evaluation is designed to improve or

---

[2]Accountability refers to evaluation conducted to demonstrate that what is evaluated is managed
with an efficient and effective use of resources and actually produces the intended benefits.

[3]The *Evaluation Exchange* is a periodical issued by Harvard University that contains new lessons
and emerging strategies for evaluating programs and policies. Since its launch in 1995, The
Evaluation Exchange has become a nationally known and significant force in helping to shape
evaluation knowledge and practice.

transform something, research is designed to prove something,[4] not supporting decision-making in understanding what is valuable. Moreover, evaluation research does not necessarily imply the use of social research methods, but it can borrow strategies, approaches, and techniques from many other fields (see e.g. Multi-Criteria Analysis or Cost-Benefit Analysis), adapting them to the necessity of drawing evaluative conclusions (Palumbo 2001).

Independently from these differences, we need to keep in mind that evaluation research is the brain of evaluation: it is the argumentation made solid, clear, replicable, verifiable because accomplished with explicit and provable processes that are recognized by the scientific-professional community, or can be inspected (Bezzi 2010). Everybody (and especially decision-makers) must be able to acknowledge the quality, the validity and the utility of information collected and its interpretation. Thanks to research the evaluation activity allows not only to determine the program effects, but also to explore theories and mechanisms that make these effects happen, during the program conceptualization and implementation. Accordingly, Weiss (1997) asserts that to understand why a program is working well or not, it is not enough to check if desired effects have been reached. On the contrary, it is necessary to go through development phases and individuate psychological, social and organizational factors that led to the results obtained, according to the circumstances and actors involved. In every situation, the link between inputs and results can be obtained in several ways, or not achieved at all.

The task of the evaluator is thus to formulate assumptions that explain why a certain intervention might lead (or not) to a certain result. This implies that what is evaluated is not given, but built by actors in a process, depending on the context and values to be measured. In every evaluation, the client, the evaluator and other stakeholders decide what to evaluate and which questions have to be answered. As a consequence, also the selection of approaches and tools to be adopted cannot be predetermined, but must be defined every time depending on the variables at hand (Stame 2001). This means that there is not a *best way* to conduct an evaluation, but every evaluation strategy needs to be designed before being executed. Before exploring how evaluation can be integrated into the service design process and how it can be conducted, some considerations are done on the concept of evaluation in the service sector and in the field of service design.

5.1.2   Evaluation in the Service Sector

Evaluation practice in the service sector is traditionally considered as a tool for managers to plan and monitor their activities for the better, and since the industrial

---

[4]Some exceptions must be recognized referring to this statement, since in some cases social research aims to enable a change too. For example, *action research* involves undertaking actions toward an improvement of the social state of affairs, and policy analysis is directly connected to decision-making (Mathison 2008).

age it is traditionally linked to management studies (Seddon 2003; Polaine et al. 2013). In that period, leading companies in the manufacturing sector, such as General Motors, began developing quantitative systems for organization evaluations, in order to support their management role. Several service companies still adopt this approach to evaluation, but being born in the industrial tradition, it comes to evidence that this method lacks a systemic vision, particularly when it comes to the service performance from the user perspective (Polaine et al. 2013).

Although a long time has passed from the industrial age, as well as from the emerging of the service economy, a comprehensive definition and understanding of service evaluation still needs to be found. In fact, both in 2010 and in 2015 Ostrom et al. identified research topics strictly connected to the understanding and measurement of service value (both from the firm and the user point of view), service performance, and impact as priorities to be addressed by *service science*.[5] The term *service evaluation* is barely used in service-focused literature.

Exploring service marketing and management literature, it can be easily realized that so far the evaluation of services has focused on service quality (Foglieni and Holmlid 2017); in a first stance concentrating on the measurement of quantitative aspects of service production to then move to the qualitative measurement of user perceptions (Grönroos 2000). Service quality can be considered the common scale for determining service value, even if, also in this case, a conventional meaning still needs to be found. This is because measuring the quality of a service mostly depends on how actors perceive it (i.e. final user, provider, designer) and, as such, it is challenging to control (Hollins et al. 2003).

Starting from the mid-1980s (see Sect. 2.3), there has been continued research on the definition, modeling, measurement, data collection procedures, data analysis, and so on of service quality (Seth et al. 2005). The most well-spread evaluation technique in the service sector, SERVQUAL (Parasuraman et al. 1988), has been developed for measuring service quality through the measurement of the compliance between customers' expectations and perceptions with respect to the service performance. The idea of comparing expectations and perceptions in order to determine service quality has strongly influenced the contemporary conception of service evaluation, as well as the conceptualization of other service evaluation techniques, such as, for example, the more recent Net Promoter Score (Reichheld 2003). On the other hand, this model has also been harshly criticized and re-formulated (Bolton and Drew 1991; Cronin and Taylor 1992), and several adaptations have been produced. The alternative version called SERVPERF (Cronin and Taylor 1992), for example, focuses on the performance component of service quality through the measurement of customer satisfaction and its impact on purchase intentions. This has contributed to making the distinction between

---

[5]*Service Science, Management, and Engineering* (SSME) is a term introduced by IBM in 2004 to describe an interdisciplinary approach for service innovation, laying on the study, design, and implementation of services systems, intended as configurations of resources (people, organization, technology and information) that interact with other service systems to create mutual value (Spohrer et al. 2008).

satisfaction and quality blurred, and the measurement of quality to become functionally similar to satisfaction measurement. The issue with service quality and satisfaction measurement is that they are mostly unidimensional (Holbrook and Corfman 1985; Zeithalm 1988) focusing exclusively on the customer point of view, thus often being inadequate to assess the overall service value. Moreover, recent developments highlight that service value is not equal to service quality or customer satisfaction (Grönroos and Voima 2013; Holmlid 2014; Arvola and Holmlid 2016) since, as highlighted by *Service Logic* (see Sect. 2.3.2), service value is always co-created by a provider and a beneficiary.

Nevertheless, some scholars have defined service quality in multidimensional terms too. Lehtinen and Lehtinen (1982) for example conceived service quality as composed by physical quality (tangible aspects of the service), interactive quality (provider–customer interaction) and corporate image (image of the provider in the mind of its current and potential clients). Also Grönroos (1984) proposed a 'Total Perceived Service Quality model' according to which the expected service performance, made of functional quality (accessibility, consumer contacts, attitudes, internal relations, behavior, service mindedness and appearance) and technical quality (knowledge, technical solutions, machines, computerized systems, employees' technical skills), is influenced by marketing activities of the service provider and external influences, such as word-of-mouth, corporate image and customer needs. In more recent studies Edvardsson (2005) suggests that service quality does not only consist of a cognitive evaluation of the service performance by the customer, but it is also affected by customer emotions.

From a design point of view, Polaine et al. (2013) propose to focus instead on the service performance during the service delivery phase by measuring the immediate experiences that service users have. This entails that all the users' interfaces, as well as back-office processes, must be taken into consideration. When these elements are not designed and managed holistically—within the same performance—he quality can vary dramatically from one touchpoint to another, in a way that can compromise both the user experience and the organization efficiency. They suggest that *inward-facing* value measurement should examine how well the service is performing for the organization, making visible variations in quality all along the different steps of the delivery process; *outward-facing* value measurement should ask how well the service is achieving the results promised to the service users, revealing variations in quality perceived in relation to the service touchpoints, channels, and tools. What can be argued is that to determine service quality both the provider and the customer perspective should be taken into consideration, since, as asserted by Polaine et al. (2013), the efficiency of the organization and the user experience are rarely contradictory forces.

Many other contributions could be analyzed to understand service quality and its measurement, but a comprehensive vision has not been developed yet. Moreover, we must consider that in the contemporary context, services are changing very fast, thus prompting service quality dimensions to evolve with them, together with ways in which they are evaluated by organizations and customers, according to the transformation of social dynamics. For example, in the last decade, digital systems have made data collection radically cheaper and more accessible both for decision-makers, people operating in the service sector and users (Hatry 2014).

Foglieni and Holmlid (2017) try to solve this uncertainty in determining service value exploring the link between value creation and service evaluation, through the analysis of *Service Logic* literature in comparison to the program evaluation literature that is also proposed in this chapter. They come up with a framework that identifies timing and perspectives of value creation in services to understand when service evaluation should be conducted and what to evaluate in order to determine specific value objectives (for the provider and the user). Accordingly, they provide a tentative definition of service evaluation as a strategy aimed at determining the value that a service has to its provider and its customer, separately with respect to service delivery and use, or jointly concerning the moments of interaction (Foglieni and Holmlid 2017). Based on this definition, which looks coherent with our vision on service design where both perspectives need to be taken into consideration (see Chap. 3), in the following sections we explore where and how service evaluation can be integrated into the service design process. But before doing that, we think it is relevant to reflect on the difference between service evaluation (as it has been defined above) and the evaluation of service design.

5.2 Evaluating Services or Evaluating (Service) Design?

According to what was discussed in Chap. 4, talking about evaluation in the service design practice does not only imply reflecting on when it should or could be conducted, but also distinguishing between the evaluation of services (or service concepts and prototypes) and the evaluation of service design.

In the design field, evaluation has been for a long time (and we would argue that it still is) connected to measuring the impact of design activities on innovation, in order to justify investments in such competencies. The measurement of design practices started in the early 2000s as demonstrated by publications such as the 'Frascati Manual' (OECD 2002) the 'Oslo Manual' (OECD and Eurostat 2005), and a research led by the University of Cambridge into the attempt of producing an 'International Design Scoreboard' (Moultrie and Livesey 2009) that compares design capabilities across nations. These reports not only contributed in understanding how innovation occurs in firms, but also led to the recognition of design as an innovation activity. Referring in particular to service design, within these studies, it is acknowledged as an essential activity for user-centered innovation. Nevertheless, there is still a lot of pressure on design to show meaningful results, both when applied to national systems, and when applied to individual firms and particular public contexts (Raulik et al. 2008). Focusing especially on service design, very few scholars and institutions have focused on how to measure its value at the organization level. This is also linked to the fact that service innovation is still a debated issue, thus making its measurement controversial.

In February 2013 the OECD promoted a workshop on measuring design and its role in innovation as part of a project carried out under the auspices of the Working

Party of National Experts on Science and Technology Indicators (NESTI), aimed at reviewing the measurement frameworks for R&D and innovation. One of the topics debated was the domain of services considered as an underdeveloped argument in the field of economics and in the measurement of innovation. The project report (Galindo-Rueda and Millot 2015) also underlines that how to deal with service innovation is a controversial point. Despite services representing an important part of the European GDP, the perception of citizens about the prominence of services in the contemporary economy seems to be low, still linking the idea of innovation to products or industrial processes. As a matter of fact, existing metrics mainly refer to a linear model of innovation[6] constraining policy-makers and investors in excluding the so-called *hidden innovation*, namely the innovation system related to services, public sector and creative industries as well as open innovation or user-centered innovation (Nesta 2008). At the public level, there is no structured evaluation strategy to support service innovation both from the organization and the user point of view. Planning and control mainly refer to financial flows and statistics that till now have driven decision-making processes. In the business sector also, in some cases, service organizations are willing to measure, but it is only when they are offered free services as part of a pilot project funded by the public sector, or pro-bono support from a private company that they are able to do it (Lyon and Arvidson 2011).

Løvlie et al. (2008) are among the few scholars who have reflected upon how to measure the value of service design practices at micro-scale. In a first stance, their idea simply consisted of working with users to assess the quality of the design work, by measuring how much people loved the designed service and were happy to use it, in agreement with traditional service quality and satisfaction measurements promoted by marketing studies. However, they soon realized that customer satisfaction only allowed to know whether people liked the service or not, not providing evidence about what is wrong at the organization level that affects the user experience. To solve this problem, they proposed to calculate the ROI of their design initiatives, in three different ways:

1. At project level, prototyping the service with a small prototype community;
2. Using a Triple Bottom Line, which measures organizational success as the sum of the economic, environmental, and social effects of an activity or, in this case, service;
3. Using a Service Usability (SU) index, which consists of a system that measures the quality of a service experience through four parameters (proposition, experience, usability, and accessibility). Data are collected qualitatively through in-depth interviews, and by shadowing users while using the service in their own environment and time.

---

[6]The linear model of innovation postulates that innovation starts with basic research, followed by applied research and development, and ends with production and diffusion. This model has been very influential and widely disseminated by academic organizations and economists as expert advisors to policy-makers, justifying its use (Godin 2006: 639–640).

What is interesting to notice is that what the authors actually propose is to evaluate the service (through prototypes) in order to support the design of the service, but they assert that what they are measuring is the value of the service design intervention, bringing about a confused vision on the difference between service evaluation and service design evaluation. We argue that these must be considered as two distinct practices. A study sponsored by the UK's Arts and Humanities Research Council (AHRC)[7] in 2014 on public and third-sector service innovation projects shows that value in design can be recognized within its cross-disciplinary approach, mindset and role and it is related to codified models (e.g. the Design Council's 'Double Diamond' model) and processes. These include the capacity of distilling and synthesizing through visualization, producing outputs that are tangible and open to critique and imagining futures through stories and artifacts.

This demonstrates that different values drive service evaluation and service design evaluation: the first one aims at assessing the quality of the service through the evaluation of the service experience (from the user perspective) and related production and delivery processes (from the provider perspective), while the second aims at assessing the quality of the service design intervention through the evaluation of design activities, competencies, and tools. The quality of a service that has undertaken a service design process could eventually, but not necessarily, be considered as an indicator of service design value, but it is not our purpose to demonstrate that. What we are going to propose is a renewed service design process that integrates (service) evaluation, in order to support the design of better services.

5.3 Rethinking the Service Design Process

Considerations about the current experience of evaluation in service design practice emerging from the analysis of case studies in Chap. 4, and the overview on the meaning of evaluation (with particular focus on the service sector) provided in this chapter make us understand that it is worth including evaluation as a systematic activity in the service design process. Referring to its capacity to drive learning and transformation, evaluation can support service design step-by-step providing evidence for decision-making, both at organization and design level. This requires rethinking the service design process, integrating evaluation within or across all of its phases. Figure 5.1 shows our proposal for this integration to happen. We can identify four evaluation stages, even though a distinction needs to be made depending on the nature of the project, namely if the purpose is to design a new service or to redesign an existing one.

---

[7]The study is entitled 'Identifying and Mapping Design Impact and Value'; see www.valuing-design.org.

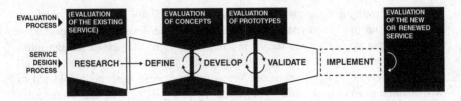

Fig. 5.1 The integrated process of service design and evaluation

In the case of service redesign only, the first evaluation phase coincides with the research phase of the traditional service design process. The focus of evaluation is on the existing service with respect to values identified as objectives of the project, to assess whether and why they are not achieved and develop solutions specifically addressed to fix these problems. When the project consists of the design of a new service, the research phase does not have an evaluation purpose, and it is traditionally conducted to understand the context, explore user needs and expectations, and so on (see Sect. 3.3.1). The successive evaluation phases (see Fig. 5.1) apply to both kinds of projects.

The second one, evaluation of concepts, and the third one, evaluation of prototypes do not imply service evaluation as it has been defined in Sect. 5.1.2 since at these stages of the service design process the service is still under definition and development. Nonetheless, as highlighted by case studies analyzed in Chap. 4, it is fundamental to monitor if the project is keeping on responding to its value objectives. It can be considered a sort of *in-itinere* evaluation, which in program evaluation literature is described as a type of assessment that accompanies the program execution, with the purpose of having continuous information upon the compliance of the program to the achievement of predefined objectives (Bezzi 2007). More precisely, the evaluation of concepts allows selecting the most promising ideas emerging from the generative sessions characterizing the *define phase* of the process. The evaluation of prototypes allows understanding what works and what does not work in the idea under development, to finally validate it. They are both iterative activities, since the results of evaluation may highlight the need to redefine the concept or to revise some elements (or to develop new ones). This is why they are positioned in-between two phases of the process.

The last evaluation phase can be assimilated to the evaluation of an existing service since it applies to the new or renewed service resulting from the implementation phase once it has been officially launched. Even though it is positioned at the end of the service design process, this evaluation should also be conducted after a longer period of time, in order to assess if the service is keeping on responding to value objectives, or if they have changed, whether and how it responds to new ones, eventually triggering (new) redesign interventions. Concerning redesign projects, evaluation of the redesigned service is fundamental to assess the value of improvements implemented compared to the initial service, while the evaluation of a new service allows understanding the value of the solution delivered.

According to this proposal of integration, a new definition of service evaluation can be formulated. In service design practice, service evaluation can be defined as *an activity aimed at determining the value of a service before and/or after the service design intervention, as well as the value of concepts and prototypes defined and developed during the service design process.* Based on this definition, and the characteristics of the different evaluation phases within the process, we can now go deep into how each phase can be conducted, or in general terms, how to design a service evaluation strategy.

5.4 Designing a Service Evaluation Strategy

Exploring the meaning of evaluation, it is possible to assume that a better way to conduct it does not exist and it is necessary to design a proper strategy every time an evaluation is required, depending on the context in which it takes place and the purposes it needs to address. The first action of the evaluator consists, therefore, in setting up the evaluation strategy, negotiating with key stakeholders the evaluation questions to be answered, the methods to be used and the relationships to be developed, also considering the resources available (Rossi et al. 2004).

Referring to program evaluation literature (see e.g. Bezzi 2007, 2010; Davidson 2005; Rossi et al. 2004; Scriven 2013), we can identify four key elements (Fig. 5.2) that should be defined by the evaluator when designing an evaluation strategy, which are also mentioned by Foglieni and Holmlid (2017) in relation to service evaluation. These key elements represent an interpretation made by the evaluator of the evaluation needs expressed by decision-makers.

- *The evaluation objectives*, that is, dimensions of merit or value (Scriven 2013) to be assessed through the evaluation process. To define evaluation objectives, Rossi et al. (2004) suggest focusing on the questions the evaluation is to answer. Evaluation questions ask if what is evaluated is valuable, addressing specific characteristics, for example: is the car safe? Is the doctor competent? Is the restaurant profitable? They can derive both from input from decision-makers and preliminary analysis as well as interpretations made by the evaluator, especially when decision-makers do not have a clear idea of the objectives they want to obtain. The formulation of specific questions allows the evaluator to understand what to evaluate to satisfy the decision-maker mandate, for example, the satisfaction of people, the effectiveness of processes, economic aspects and so on. Furthermore, they facilitate the selection of proper data collection method and techniques and address the interpretation of results.
  Concerning the definition of evaluation objectives, an important aspect needs to be taken into account: their *evaluability*, namely if it is possible to evaluate what the decision-maker would like to evaluate. Palumbo (2001) asserts that everything is evaluable, but many problems can arise when facing the evaluation design, such as conflicts between stakeholders, unrealistic objectives, lack of

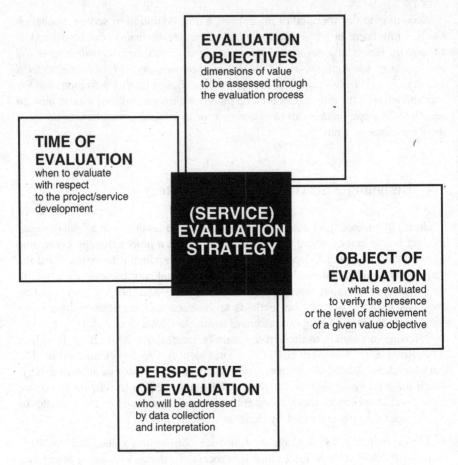

Fig. 5.2 Key elements to design a (service) evaluation strategy

resources, and so on. According to Wholey (1994) the minimum conditions for
something to be evaluable are:

- The objectives must be plausible and shareable/shared among the
  stakeholders;
- Information about what is evaluated must be accessible to the evaluator;
- There must be a shared vision of how to use the evaluation results among
  stakeholders involved.

- *The object of evaluation*, that is, what is evaluated, in terms of processes,
  performances, organizations, products, and so on, to verify the presence or the
  level of achievement of a given value objective. Defining what to evaluate is
  strictly related to the definition of value objectives and, consequently, needs to
  be negotiated with stakeholders. Once it has been defined, it is also necessary to
  set indicators of merit or value (Scriven 2013) that establish the status under

which or over which what is evaluated is considered achieving (or not) value objectives. Indicators can be defined as observable levels or quantities that allow describing something (Bezzi et al. 2010; Marradi 1987). In this sense, the possible indicators are infinite. They can be given or built ad hoc, but always in accordance with the value system of the evaluation at hand.

- *The perspective of evaluation*, that is, who will be addressed by data collection and interpretation. To frame the value system of an evaluation, stakeholders and their individual value systems play a key role. Thus far, we mainly mentioned the evaluator and decision-makers as key actors of the process, since they guide the negotiation about what and how to evaluate. However, although not directly involved in the evaluation design, they also target participants/users, managers, and staff, and directly affect the definition of value objectives and related evaluation objects. As a matter of fact, operators and beneficiaries of the program to be evaluated are usually at the core of data collection and interpretation, determining the evaluation perspective implied in the evaluation object. Scriven (2013) defines them as *overt audience*, intending people directly addressed by the assessment or who have the right to know its results.
- *The time of evaluation*, that is, when to evaluate the program with respect to its design, implementation, and use. Considering programs (but also services) as sequences of activities that happen over a period of time, determining the time of evaluation helps define, for example, whether the objective of the evaluation is the efficiency of activities held by the staff, or the impact on target users once the program is concluded (Altieri 2009; Bertin 1995; Palumbo 2001). Scholars usually distinguish between (i) evaluations conducted in predictive terms (*ex-ante*) to assess the enabling conditions for a program, activity or service to be developed; (ii) evaluations aimed at monitoring the development of a program, activity, or service execution; and (iii) evaluations conducted once the program, activity, or service is concluded (*ex-post*) to verify its planned and unplanned outputs. In reality, it is not necessary to define a chronological order, since it is not always possible to evaluate all the aspects of a program, thus making each phase independent from the others and characterized by specific objectives, objects, and perspectives (Palumbo 2001).

How do these elements apply to the evaluation of services? If we substitute the word *program* with the word *service* it becomes easier to imagine how to translate these elements to the design of a service evaluation strategy. Nonetheless, it is worth further exploring each element to better understand how to pragmatically integrate evaluation into the service design practice.

In the previous section, we have analyzed the positioning of evaluation in the service design process, which allowed rethinking the process itself. Considering the process as a sequence of activities that happen over time, it is evident how the evaluation phases identified also acquire a temporal dimension. This determines the possible times of evaluation regarding the development of a service design project and, as a consequence, influences the definition of evaluation objectives and objects that are specific for each moment of development. If we consider, instead, the evaluation of an existing service or a (re)new(ed) service as the result of a service design project, further possible evaluation moments can be identified, which concern the process of value creation (cf. e.g. Grönroos and Voima 2013) as suggested by Foglieni and Holmlid (2017). These can coincide, for example, with the moment of service production (before the service is delivered), the time of delivery and use, the post-purchase experience, and so on.

For what concerns the evaluation perspective, when we evaluate a service we need to consider, at least, two perspectives: that of the service provider and that of the service user, as emerging from the definition formulated in the previous section and from the vision of Polaine et al. (2013) about *inward-facing* and *outward-facing* value measurement (see Sect. 5.1.2). Evaluation is done from the provider perspective when the object of assessment investigates organization processes, performances and resources (Manschot and Sleeswijk Visser 2011) and related value objectives. It is done from the user perspective when the object of evaluation examines customer expectations and perceptions before, during or after the use of the service (Heinonen et al. 2010, 2013; Helkkula and Kelleher 2010). Referring to the value creation process described in *Service Logic* literature, Foglieni and Holmlid (2017) identify a further perspective that should be taken into consideration when designing a service evaluation strategy: the *joint perspective* (Grönroos and Ravald 2011; Grönroos and Voima 2013), concerning the moments of direct and indirect interactions between the provider and the user that occur at the intersection of the delivery and usage phases. However, we can consider the *joint perspective* as part of other perspectives, since also the evaluation of interactions needs to be addressed toward provider or user values.

Differently from the time and the perspective of evaluation, understanding what to evaluate in a service cannot be predetermined. Objectives and objects of evaluation are at the core of the negotiation process with those who commission the evaluation, and in the case of an evaluation integrated into a service design project, we argue they should also be coherent with the design aims of the project. Program evaluation literature (e.g. Bezzi 2007; Scriven 2013) vaguely indicates as possible objects of evaluation physical standards, individual performances or habitual patterns of behaviors, processes, and organizations. This indication can also apply to services, being considered as complex social systems (Spohrer et al. 2008; Vargo and Lusch 2014) that might include products, people, technologies, information, processes, and experiences. Another useful categorization for the definition of evaluation objects is that provided by Vargo and Lusch (2014) about resources, intended as physical artifacts, systems, skills, and knowledge possessed, produced and exchanged by the service provider and the service user. They distinguish between *operand resources*, which are static and require some action to be performed on them to provide value (like products), and *operant resources* capable of acting on other resources to create value, like human competencies and knowledge. Accordingly, in the attempt to identify what can be evaluated in a service, Foglieni and Holmlid (2017) distinguish between *service elements*, intended as the tangible aspects of the service (such as touchpoints and interfaces) and resources necessary to produce, access, and use the service; and *processes*, intended as sequences of activities aimed at producing, delivering, or experiencing the service. More precisely, they argue that when evaluation is conducted from the provider perspective, it should focus on resources possessed or prepared by the provider in planning and producing organization processes, back-office processes necessary to deliver the service and their outputs and outcomes. When it is conducted from the user perspective it should focus instead on expectations connected to inputs from the provider and the user social context, and perceptions and memory of the service experience. Moreover, during interactions they suggest evaluation should focus on the usage of resources, the service delivery and use processes (front-office processes), and experiences of users and staff.

As stated before, the evaluation of these objects allows determining the presence or the level of achievement of predefined value dimensions that represent the objectives of evaluation, so to eventually undertake actions that can introduce or better address them. Similarly to objects, it is tough to prefigure value objectives for service evaluation. Some common service values that are usually addressed by evaluation are service quality and customer satisfaction, as discussed in Sect. 5.1.2. But they can also be more specific applying to a single service element or to the characteristics of a given sector as emerging from the analysis of case studies in Chap. 4, where evaluation focuses for example on the usability of digital interfaces or patient safety. A common way to define value objectives for organizations is to individuate KPIs (Manschot and Sleeswijk Visser 2011), but in our opinion this is not mandatory, especially considering value objectives related to the user perspective.

To conclude, we want to focus on one aspect that must always be taken into consideration when designing a (service) evaluation strategy: the problem of *objectivity* and *subjectivity*. As outlined in the first chapter, what distinguishes informal and formal evaluations is that the former are highly personal and subjective, while the latter aim at producing a scientific or objective knowledge. In reality, since evaluation is strongly context-dependent and reliant on interpretations of the evaluator both concerning how to design the evaluation and the formulation of evaluative conclusions emerging from it, a high degree of subjectivity is always involved (Bezzi 2007, 2010; Davidson 2005). This implies a significant responsibility for the evaluator, who thus needs to build a solid argumentation that supports each of his/her choices about the core elements of evaluation described in this section. Starting from the design of the evaluation strategy, in the next section, some guidelines are proposed to transform the strategy into a pragmatic evaluation process that can accompany the service design process.

5.5 Guidelines for Service Evaluation and Design

In this chapter we have learnt what it means to evaluate a service, when to conduct evaluation during the service design process and how to set up a service evaluation strategy. In this final section, some practical steps that can support the service designer in kick-starting an evaluation process within a service design project are proposed. We refer once again to program evaluation literature, which offers good indications that are transversal to any evaluation context.

The design process of evaluation is a matter of *problematization–argumentation–reflection–interpretation*. It is, first of all, a mental process leading the evaluator to concretely realize the evaluation research, from the definition of the initial cognitive problem that constitutes his or her mandate and justifying every methodological choice (Carvalho and White 2004). On the one hand, it allows framing the context, the logic of stakeholders and their value systems, and analyzes the preconditions for and implications of conducting the evaluation. On the other hand, it allows defining the research strategy, selecting data collection and interpretation methods and techniques and specifying how to use them, to formulate evaluative conclusions (Bezzi 2010). Scholars (see e.g. Bezzi 2007, 2010; Davidson 2005; Scriven 2013) describe a set of steps or phases that should guide the design process of an evaluation strategy. They are proposed as a sequence of actions, but it must be considered that being a design process, it is iterative and all the actions can be repeated and revised according to emerging issues. They can be clustered into three main groups:

1. A preliminary group of phases aimed at collecting basic information about the evaluation and framing the context, including the acquisition of the evaluation mandate, the definition of value objectives and evaluation questions, the check of available resources;
2. A second group of phases aimed at defining the research strategy to be conducted, including evaluation objects, times and perspectives, and related indicators, as well as the selection of data collection and interpretation methods and techniques;
3. A third group of phases aimed at operationalizing the research, analyzing results and building conclusions to propose improvement or transformation recommendations.

These phases can also be applied to service evaluation. In particular, we suggest the following guidelines for developing a service design project that also includes evaluation. They are presented in a logical order, but some steps can be iterated, as shown in Fig. 5.3.

1. **Learn about the context**: immersion in the service context to understand why a service design intervention is needed. When a service organization decides to design a new service or redesign the service it provides, it is usually because something is going wrong, such as a decrease in profits, or new requirements are emerging regarding user needs, technological changes, availability of resources, and so on. Recognizing this could just be based on a feeling, or could emerge from internal (partial) evaluations and market research conducted by the organization. People in charge of service design and evaluation thus need to have a clear understanding of the starting situation and the overall transformations characterizing the context, establishing a dialogue with decision-makers and

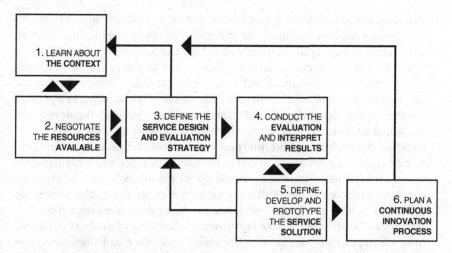

Fig. 5.3 Guidelines for service evaluation and design

eventually other stakeholders involved (including users if needed), entering into their way of thinking and valuing.

2. **Negotiate the resources available**: ascertain the human, temporal, and economic resources available for the entire project and distribute them through the activities involved in the different steps. Evaluation is often perceived in negative terms and tends to be considered a worthless and resource-consuming activity within organizations. Unless it is compulsorily required by investors (especially in the public sector) for accountability purposes, it is unusual for an organization to envisage dedicating resources to evaluation in a service design project. People in charge of service design and evaluation thus need to understand and distribute in a smart way resources available for the entire project, explicitly allocating some of them to evaluation and others to design. The resources required consist of money, time, and people. In the case of a scarcity of resources, the task of the service designer is also to determine creative ways of overcoming limitations and constraints or to push decision-makers into investing more resources. A secondary objective here is also to educate service organizations concerning the importance of service design and evaluation to enable a continuous innovation process.

3. **Define the service design and evaluation strategy**: structure design activities and decide when to conduct the evaluation during the service design process; define the value system based on which designs and evaluations are conducted. Depending on the context and resources available people in charge of service design and evaluation need to structure the integrated service design and evaluation process. Concerning evaluation, in the case of a redesign project, an evaluation of the initial service should be included, while in the case of the design of a new service evaluation could focus on concepts and prototypes. In both cases, the evaluation of the new or renewed service resulting from the project

should be conducted. Starting from the formulation of evaluation questions, for each evaluation phase specific value objectives and objects are identified, as well as the evaluation perspective to be undertaken, and tools to be adopted for data collection and interpretation. Accordingly, activities for each phase of the service design process are structured and scheduled over time, including purposes, expected outputs, and tools. At this point, the service designer needs to recognize whether or not he or she has sufficient expertise to conduct the design and evaluation activities and decide whether to involve other competencies.

4. **Conduct the evaluation and interpret results**: undertake data collection concerning the service organization and/or the user experience and formulate evaluative conclusions that identify problematic and opportunistic areas to be addressed by service design. Following the service evaluation strategy previously defined, it is now possible to operationalize the evaluation research. Data collection can focus on the service organization when the evaluation is conducted from the organization perspective, including back-office and front-office processes, touchpoints and interactions characterizing service execution and delivery. Or it can focus on the user experience when the evaluation is conducted from the user perspective, including the perceptions of users while using the service and based on their memories of the service, but also expectations arising from use of the service, according to the users' individual and collective contexts. Evaluation results are interpreted to frame critical points and finding reasons why given values are not being satisfied. Once the evaluation has been conducted, a clear picture of the service, concept or prototype both from the organization and the user perspectives will be afforded the service designer, allowing him/her to establish design directions to be followed, or the form of the change(s) required.

5. **Define, develop and prototype the service solution**: engage in concept generation, select of the most promising ideas, develop and test them through prototyping before undertaking a full implementation of the service. Evaluation results give evidence to every choice and solidity to every solution proposed. Based on these results the service designer develops the service following the traditional phases of the service design process, involving the organization and users in co-design activities when necessary and iterating some steps if required, up to the definition of the final solution.

6. **Plan a continuous innovation process**: develop a plan for embedding a continuous innovation process supported by service evaluation and design, aimed at educating the organization in a transformative culture. It is much harder to undertake a significant change once in awhile than periodic small improvements. To do so and keep the organization competitive and the service innovative, it is necessary to endow the organization with evaluation and design capabilities. Once the service has been implemented, a plan needs to be developed to monitor how it performs and how users' perceptions evolve, enabling continuous updating of the value system, and assess the extent to which the service responds.

As it can be argued from our guidelines designing an evaluation strategy is not that different from designing any other thing. It is a process made up of steps that require the adoption of certain tools. Speaking of which, it is worth reflecting on what kinds of tools can accompany the execution of such integrated evaluation and design processes. We want to focus in particular on tools to be adopted for evaluation since tools to be used during each step of the service design process have already been mentioned in Chap. 3. The analysis of case studies has shown that some service design tools typically adopted during the research phase (see Sect. 3.3.1) can also be used for evaluation purposes. Moreover, we must not forget that service design tools are endowed with a robust visual capacity, which supports interpretation stages of the creation process (Segelström 2009). Concerning service evaluation, visualization skills of service designers can thus represent an added value for data interpretation, because it is a way to make insights shareable and available to critique. From an evaluation perspective, it can be argued that visualization tools typically used in service design can, on one hand, support the visualization of evaluation results, and on the other hand facilitate interpretation through a visual representation of collected data. Nonetheless, using tools commonly adopted in service design practice for data collection and interpretation can make evaluation easier for service design practitioners. What is worth highlighting is that applying tools is not enough either for evaluation or for service design: what is needed is a strategy considering the context in which the project takes place and its purposes in terms of value. Only by developing a strategy aimed at defining what is being evaluated and designed, for which reasons and through which process, will it be possible to select appropriate tools and obtain consistent results.

References

Altieri L (2009) Valutazione e partecipazione. Metodologia per una ricerca interattiva e negoziale. FrancoAngeli, Milano

Arvola M, Holmlid S (2016) Service design ways to value-in-use. In: Proceedings of ServDes 2016, Aalborg University, Copenhagen, 24–26 May 2016

Bertin G (1995) Valutazione e sapere sociologico. Franco Angeli, Milano

Bezzi C (2007) Cos'è la valutazione. Un'introduzione ai concetti, le parole chiave e i problemi metodologici. Franco Angeli, Milano

Bezzi C (2010) Il nuovo disegno della ricerca valutativa. Franco Angeli, Milano

Bezzi C, Cannavò L, Palumbo M (2010) Costruire e usare indicatori nella ricerca sociale e nella valutazione. FrancoAngeli, Milano

Bolton RN, Drew JH (1991) A multistage model of customers' assessments of service quality and value. J Consum Res 17(4):375–384

Carr VL, Sangiorgi D, Buscher M, Junginger S, Cooper R (2011) Integrating evidence-based design and experience-based approaches in healthcare service design. Health Environ Res Des J 4(4):12–33

Carvalho S, White H (2004) Theory-based evaluation: The case of social funds. Am J Eval 25 (2):141–160

Chelimsky E (2006) The purpose of evaluation in a democratic society. In: Shaw IF, Greene JC, Melvin MM (eds) The Sage handbook of evaluation. Sage, London, pp 33–55

Coffman J (2003–2004). Michael Scriven on the differences between evaluation and social research. Eval Exch 9(4):7

Cronin JJ, Taylor SA (1992) Measuring service quality: a reexamination and extension. J Mark 56(3):55–68

Davidson EJ (2005) Evaluation methodology basics: the nuts and bolts of sound evaluation. Sage, Thousand Oaks

De Ambrogio U (2003) Valutare gli interventi e le politiche sociali. Carocci Faber, Roma

Donaldson SI, Lipsey MW (2006) Roles for theory in contemporary evaluation practice: Developing practical knowledge. In: Shaw I, Greene J, Mark M (eds) Handbook of evaluation. Sage, Thousand Oaks, pp 56–75

Edvardsson B (2005) Service quality: Beyond cognitive assessment. Managing Serv Qual 15(2):127–131

Feinstein O, Beck T (2006) Evaluation of development interventions and humanitarian action. In: Shaw IF, Greene JC, Melvin MM (eds) The Sage handbook of evaluation. Sage, London, pp 536–558

Foglieni F, Holmlid S (2017) Determining service value: exploring the link between value creation and service evaluation. Serv Sci 9(1):74–90

Foglieni F, Maffei S, Villari B (2014) A research framework for service evaluation. In: Proceedings of ServDes 2014, Imagination Lancaster, Lancaster, 9–11 Apr 2014

Forss K, Marra M, Schwartz R (2011) Evaluating the complex: attribution, contribution, and beyond. Transaction Publishers, New Brunswick

Fournier D (2005) Evaluation. In: Mathison S (ed) Encyclopedia of evaluation. Sage, Thousand Oaks, pp 140–141

Galindo-Rueda F, Millot V (2015) Measuring design and its role in innovation. OECD Science, Technology and Industry Working Papers, No. 2015/01, OECD Publishing, Paris

Gamal D (2011) How to measure organization innovativeness? An overview of innovation measurement frameworks and innovative audit/management tools. http://www.tiec.gov.eg/backend/Reports/MeasuringOrganizationInnovativeness.pdf. Accessed 5 Jul 2016

Gillinson S, Horne M, Baeck P (2010) Radical efficiency—Different, better, lower cost public services. Available via http://innovationunit.org/sites/default/files/radical-efficiency180610.pdf. Accessed 23 Jan 2016

Godin B (2006) The linear model of innovation the historical construction of an analytical framework. Sci Technol Human Values 31(6):639–667

Grietens H (2008) Quali prospettive europee a proposito di interventi evidence based per bambini e ragazzi a rischio e le loro famiglie. In: Canali C, Vecchiato T, Whittaker JK (eds) Conoscere i bisogni e valutare l'efficacia degli interventi per bambini, ragazzi e famiglie in difficoltà. Fondazione Zancan, Padova

Grönroos C (1984) A service quality model and its marketing implications. Eur J Mark 18(4): 36–44

Grönroos C (2000) Service management and marketing: a customer relationship management approach, 2nd edn. John Wiley & Sons, New York

Grönroos C, Ravald A (2011) Service as business logic: Implications for value creation and marketing. J Serv Man 22(1):5–22

Grönroos C, Voima P (2013) Critical service logic: Making sense of value creation and co-creation. J Acad Mark Sci 41(2):133–150

Hammersley M (2003) Social research today: some dilemmas and distinctions. Qual Soc Work 2(1):25–44

Hatry HP (2014) Transforming Performance Measurement for the 21st Century. Washington, DC: The Urban Institute. http://www.urban.org/research/publication/transforming-performance-measurement-21st-century/view/full_report. Accessed 3 Jul 2016

Heinonen K, Strandvik T, Voima P (2013) Customer dominant value formation in service. Eur Bus Rev 25(2):104–123

Heinonen K, Strandvik T, Mickelsson KJ, Edvardsson B, Sundström E, Andresson P (2010) A customer-dominant logic of service. J Serv Man 21(4):531–548

Helkkula A, Kelleher C (2010) Circularity of customer service experience and customer perceived value. J Consumer Behav 9(1):37–53

Holbrook MB, Corfman KP (1985) Quality and value in the consumption experience: Phaedrus rides again. Perceived Quality 31(2):31–57

Hollins B, Blackman C, Shinkins S (2003). Design and its management in the service sector— Updating the standard. http://www.ub.edu/5ead/PDF/14/Hollins2.pdf. Accessed 10 Jan 2016

Holmlid S (2014) One approach to understand design's value under a service logic. In: Proceedings of the 19th DMI: Academic Design Management Conference, London, 2–4 Sep 2014

Lehtinen U, Lehtinen JR (1982) Service quality: A study of quality dimensions. Service Management Institute, Helsinki

Løvlie L, Downs C, Reason B (2008) Bottom-line experiences: measuring the value of design in service. Des Manag Rev 19(1):73–79

Lyon F, Arvidson M (2011) Social impact measurement as an entrepreneurial process. http://www.birmingham.ac.uk/generic/tsrc/documents/tsrc/working-papers/briefing-paper-66.pdf. Accessed 10 Mar 2017

Manschot M, Sleeswijk Visser F (2011) Experience-value: A framework for determining values in service design approaches. In: Proceedings of IASDR 2011, Delft, 31 Oct–4 Nov 2011

Marradi A (1987) Concetti e metodi per la ricerca sociale. La Giuntina, Firenze

Mathison S (2005) Encyclopedia of evaluation. Sage, Thousand Oaks

Mathison S (2008) What is the difference between evaluation and research—and why do we care. In: Smith L, Brandon PR (eds) Fundamental issues in evaluation. The Guildford Press, New York, pp 183–196

McNabola A, Moseley J, Reed B, Bisgaard T, Jossiasen AD, Melander C, Whicher A, Hytönen J, Schultz O (2013) Design for public good. Available via Design Council. http://www.designcouncil.org.uk/sites/default/files/asset/document/Design%20for%20Public%20Good.pdf. Accessed 15 May 2017

Morrison KRB (1993) Planning and accomplishing school centred evaluation. Peter Francis Publishers, Norfolk

Moultrie J, Livesey F (2009) International design scoreboard: initial indicators of international design capabilities. http://www.idi-design.ie/content/files/InternationalDesignScoreboard.pdf. Accessed 3 Oct 2015

Nesta (2008) Measuring innovation. Available via Nesta. https://www.nesta.org.uk/sites/default/files/measuring_innovation.pdf. Accessed 25 Feb 2016

OECD (2002) Frascati manual 2002: Proposed standard practice for surveys on research and experimental development, The measurement of scientific and technological activities. OECD Publishing, Paris

OECD (2011) Making the most of public investment in a tight fiscal environment. OECD Publishing, Paris

OECD and Eurostat (2005) Oslo Manual: Guidelines for collecting and interpreting innovation data, 3rd eds. OECD Publishing, Paris

Ostrom AL, Parasuraman A, Bowen DE, Patricio L, Voss CA (2015) Service research priorities in a rapidly changing context. J Serv Res 18(2):127–159

Palumbo M (2001) Il processo di valutazione. Decidere, programmare, valutare. Franco Angeli, Milano

Parasuraman A, Zeithaml VA, Berry LL (1988) SERVQUAL: A multiple-item scale for measuring consumer perceptions of service quality. J Retail 64:12–40

Patton MQ (1997) Utilization-focused evaluation: The new century text, 3rd edn. Sage, Thousand Oaks

Polaine A, Løvlie L, Reason B (2013) Service design: from insight to implementation. Rosenfeld Media, New York

Raulik G, Cawood G, Larsen P (2008) National design strategies and country competitive economic advantage. Des J 11(2):119–135

Reichheld FF (2003) The one number you need to grow. Harvard Bus Rev 81(12):46–55

Rossi PH, Freeman HE, Lipsey MW (2004) Evaluation. A systematic approach, 7th edn. Sage, Thousand Oacks

Scriven M (1991) Evaluation thesaurus, 4th edn. Sage, Newbury Park

Scriven M (2007) Key evaluation checklist. http://citeseerx.ist.psu.edu/viewdoc/download?doi= 10.1.1.116.2048&rep=rep1&type=pdf. Accessed 10 Nov 2015

Scriven M (2013) Key evaluation checklist. http://www.michaelscriven.info/images/KEC_3.22. 2013.pdf. Accessed 3 Dec 2016

Seddon J (2003) Freedom from command and control: a better way to make the work work. Vanguard Consulting Ltd, Buckingham

Segelström F (2009) Communicating through visualizations: service designers on visualizing user research. In: Proceedings of ServDes 2009, Oslo, 24–26 Nov 2009

Seth N, Deshmukh SG, Vrat P (2005) Service quality models: a review. Int J Qual Reliab Manag 22(9):913–949

Shaw IF, Greene JC, Melvin MM (2006) The Sage handbook of evaluation. Sage, London

Spohrer J, Vargo SL, Caswell N, Maglio PP (2008) The service system is the basic abstraction of service science. In: Proceedings of the 41st Annual Hawaii International Conference on System Sciences, Waikoloa, 7–10 Jan 2008

Stame N (2001) Tre approcci principali alla valutazione: Distinguere e combinare. In: Palumbo M (ed) Il processo di valutazione. Decidere, programmare, valutare. Franco Angeli, Milano, p 21–46

Vargo SL, Lusch RF (2014) Service-dominant logic: premises, perspectives, possibilities. Cambridge University Press, Cambridge

Weiss CH (1997) Theory-based evaluation: past, present and future. New Dir Eval 76:41–55

Wholey JS (1994) Assessing the feasibility and likely usefulness of evaluation. In: Wholey JS, Hatry HJ, Newcomer KE (eds) Handbook of practical program evaluation. Jossey-Bass, San Francisco, p 15–39

Woodruff RB, Gardial SF (1996) Know your customers—New approaches to understanding customer value and satisfaction. Blackwell, Oxford

Zeithaml VA (1988) Consumer perceptions of price, quality, and value: a means-end model and

Chapter 6
Service Design and Service Evaluation: Challenges and Future Directions

Abstract In response to challenging contemporary transformations that inevitably affect both the public and the business sectors, service design is increasingly acquiring a dominant role. Nonetheless, the debate about how to measure its actual contribution in triggering changes, and innovation to the largest extent, is still on the cutting-edge. What we argue in this chapter is that including the evaluation of services and service solutions into the service design practice, as extensively discussed in this book, can contribute to determine the value of service design itself as an effective and profitable approach that generates a positive impact within organizations. This enables an interpretation of the possible levels of adoption of evaluation in service design, from situations where it is not included at all, up to become a strategy fully embedded into the service design process. To support such levels of integration new competencies are required for service design professionals, to set up the evaluation strategy and to eventually conduct the evaluation research when professional evaluators cannot be included in the project. To conclude, some open issues are explored concerning how these issues can influence the evolution of the discipline and future challenges posed to service designers.

Keywords Service design · Service evaluation · Service design evaluation · Service value · Design value · Service design value · Adoption of evaluation · Service design competencies · Evaluation competencies

6.1 Building the Future Service Economy

The European Commission states that in the last two decades services have become a driver of growth in developed economies, resulting in net job gain (European Commission 2009). Accordingly, the importance of designing them and evaluating their effectiveness is becoming crucial to producing better solutions and performances both at the public and private level. Service providers have to face rapid changes due to quick transformations, scarcity of resources, new job models, and increasingly competitive markets, while users want to live better and have

© The Author(s) 2018
F. Foglieni et al., *Designing Better Services*, PoliMI SpringerBriefs,
https://doi.org/10.1007/978-3-319-63179-0_6

sustainable experiences from an environmental, social and economical point of view. As mentioned in Chap. 2, the European Commission describes the importance of design as a lever for growth, and services as an essential asset to foster innovation (European Commission 2009, 2013). Designing or redesigning services entails different activities that impact on users through the transformation of public administrations and companies such as ideating new service offerings, changing the modalities of interaction between people and organizations, creating or improving new business models. In this context, service design and *design for service* approaches are being discussed and validated within academic and professional communities. So far, discussions have focused on making clear the value of the design approach in fostering service innovation processes, reflecting on both new theoretical assumptions and professional practices. On the one hand, the strengthening of the discipline and the creation of a common language within the community is made evident through the diffusion of specific and well-established tools among researchers and practitioners. On the other hand, some recent reflections on the evolution of service design in terms of design objects (Kimbell and Blomberg 2017), new service development (Holmlid et al. 2017) and new fields of application such as *design for policy* (Bason 2010; Buchanan et al. 2017; Junginger 2013; Kimbell 2015, 2016) are being reinforced.

As outlined in Chap. 3, the service design process involves peculiar skills and contributions from different disciplines, and it is divided into a sequence of steps that include research, design, and development of solutions that revolve around user needs and organizational goals. From our point of view, we need to further reflect on how to exploit the unexpressed potential of service design and understand what areas need to be deepened. The hypothesis that guides this book is that the evaluation of services is a significant element that should be introduced as a disciplinary component, first to understand and measure service value and second to understand and measure service design value. As stated before, someone already started a reflection on this topic, but it remains underexplored. We have therefore introduced the subject of evaluation as an element of innovation within the service design process used to guide and assess the results of service design interventions.

To understand what is the current role of evaluation in the design process of services we analyzed some case studies developed in different fields (see Chap. 4) to understand how evaluation is perceived, designed and operationalized within organizations and consultancy agencies. Then, based on the exploration of the topic in extra-disciplinary fields of study, and paying attention to the difference between assessing services and measuring the value of service design, we reflected on how to pragmatically include evaluation in service design practice, proposing an integrated service design and evaluation process supported by a set of guidelines. In this concluding chapter, we explore the challenges that such a novel approach poses to the discipline, including new competencies required for practitioners and open issues for the next future.

6.2 Measuring the Value of Service Design: A Future Challenge for Service Designers

The analysis of literature and examples investigated demonstrate how it is still difficult to find an acknowledged process to evaluate services through the service design lenses. In some cases, evaluation can be considered as a research process aimed at determining the value of services and service solutions; in other situations, it can be adopted to measure the capacity of service design to create value for the organization and contribute to innovation. In the previous chapter, we focused on service evaluation as a support practice for service design practice, while referring to the measurement of the value of the service design approach further considerations need to be undertaken.

As reported in Sect. 4.1.2, the discourse on how to evaluate services finds its origins in dedicated disciplines such as service marketing and management, and it is mainly connected to the measurement of service quality and customer satisfaction in relation to organizational performances. How these kinds of measures could apply to service design to assess its value, or how service design can enhance the achievement of such measures are poorly explored topics both in service marketing (Andreassen et al. 2016) and in service design literature. Because of its characteristics (see Chap. 3), the link between the adoption of service design and the development of useful and successful services is often regarded as a causal relationship. In fact, this is far from being demonstrated since it would require to monitor service design solutions for a reasonable period after they have been implemented and to involve expertise on impact measurement, which is much more complicated than the kind of evaluation we have dealt with in this book. *Impact evaluation* should be conducted at least 12–18 months after the service launch and it is a complex process that cannot be made with certainty, but only with varying degrees of confidence. *Impacts* refer to that portion of an *outcome*[1] that can be attributed uniquely to a program or service, with the influence of other sources, the so-called *confounding factors*, controlled or removed (e.g. variation in the employment rate in the target population of the area where an employment service has been delivered) (Rossi et al. 2004). Determining *confounding factors* often requires a parallel research (on the evaluation research, see Sect. 4.1.1) to be developed (Palumbo 2001). Moreover, the target population might be difficult to reach, or it might be difficult to obtain follow-up data from them (Rossi et al. 2004),

---

[1]According to program evaluation literature *ex-post* evaluations can address *outputs, outcomes* or *impacts*. *Outputs* are defined as what the program or service concretely realizes, its physical evidences or internal efficacy, i.e. the capacity of achieving predefined objectives (Bezzi 2010; Palumbo 2001). They are the immediate and programmable consequences (e.g. the number of people attending a course). *Outcomes* are the results of the program or service, its effective and not necessarily planned consequences (e.g. number of people who found a job after attending the course) or external efficacy, i.e. the capacity to produce benefits for its beneficiaries (Palumbo 2001).

without considering the amount of economic, human and temporal resources that whoever commissions the evaluation should make available.

Thus, what we are wondering is, is it possible to establish a causal relationship between a service success and the adoption of service design? Is it possible to measure the contribution of service design to service innovation? What is the impact of service design, from a social and monetary perspective (and not only that), and how can we evaluate it? In the attempt to answer these questions, we refer to studies about the measurement of design value, which in the academic field is a more mature topic that could also partially apply to service design, despite it remaining a critical issue. We report on the work done by some authors mainly referring to the design management literature. Although it has been shown that design brings value to organizations at the economic, strategic and social levels (Borja de Mozota 2002, 2011; Manzini 2014; Micheli 2014; Thomson and Koskinen 2012; Villari 2012), it often remains confined to a marginal role, since it is seen as a mere function or tool applied with various modalities within organizations (Junginger 2009). Similarly, according to Borja de Mozota (2002), design produces value through a *sensemaking attitude*, defining aesthetic features of products and user analysis processes.

On the European scale, design is measured mainly as a component of innovation processes at the firm level (European Commission 2013; Moultrie and Livesey 2009). One of the most well-known studies about the measurement of design value is the 'Design Ladder' (Ramlau and Melander 2004), which is used to estimate the design maturity level in enterprises. It proposes a hierarchical model divided into four phases, from an initial level in which design is not used and has no importance for the company up to the fourth level (the highest grade on the scale) in which design is an integral part of innovation processes and blended into the top management. The 'Design Ladder' describes degrees of growing complexity (no design, design as styling, design as a process, and design as a strategy) to outline different ways through which firms explicitly adopt design for their operations.

Within the field of design management, the 'Design Management Staircase' (Kootstra 2009) also was developed as an instrument for self-assessment dedicated to organizations and referred to the new product development framework. In this case, the scale describes four degrees of maturity closely linked to the use of design management in companies (no design management, design management as a project, design management as a function, design management as a culture). Similarly, the Design Management Institute (DMI) has published a series of reports and tools called 'The Design Value System'. These aim at communicating the value of investment in design, assessing the maturity of design organizations, and benchmarking the firm's domains in which design adds value.

Concerning the topic of investments, the DMI's 'Design-Centric Index' (Westcott et al. 2013) maps the best methods and metrics for measuring and managing design investments. The authors define three key elements to describe how organizations use design: (i) as a service, mainly considered in its aesthetic and functional dimensions with a demonstrable ROI impact; (ii) as connector or integrator when businesses are more related to user-centered approaches rather than departmental and product focused; and (iii) as a strategic resource when companies use design as a strategic lever and as a core competence. Accordingly, different metrics can be identified for design assessment: qualitative metrics generally consist of ROI or profit margins and stock performance; qualitative metrics consider for example the lifetime customer value or brand loyalty.

Another model for measuring the value of design in organizations is described through the concept of design capability (Mortati et al. 2014; Mortati and Villari 2016). According to this model, the impact of design processes on business performances is assessed through three design capabilities: design leadership, design management, and design execution intended as different ways to manage, integrate, access or introduce internal and external design resources within the organization. Moreover, the €Design project[2] (co-funded by the European Commission Directorate General Enterprise & Industry) has created a set of tools to measure the positive impact of design, to encourage policy-makers and businesses across Europe to integrate design into their work. In particular, a communication toolkit is proposed to help organizations developing appropriate skills to measure and manage design innovation in businesses.

Within the public sector, the measurement of the design value looks more oriented toward service design. A similar process to the 'Design Ladder' has been described in the report 'Design for public good' (McNabola et al. 2013). It defines a qualitative measurement approach through examples that describe the level of maturity of governments in using (service) design to promote innovation. A 'Public Sector Design Ladder' is proposed. It is based on three steps (design for discrete problems, design as capability, design for policy) that indicate how public organizations and decision-makers use design as a lever for innovation. The first level describes the situation in which design is not embedded in the public organization, and it is used to make technology user-friendly. The second step considers those cases in which designers and public sector employees work together and use design tools and techniques. The last step refers to those contexts in which design is adopted at all levels of the public organization and used by policy-makers to support their initiatives.

Looking specifically at the measurement of service design value, Manschot and Sleeswijk Visser (2011) have proposed a framework for the assessment of service design through the evaluation of services, which in their opinion is based on the perceptions of people while using a service. They identify two types of value: the value of system performance (attributed to the organization) measured through performance indicators, and the value of personal experiences (of service users). This vision is similar to that developed by Løvlie et al. (2008) and described in Sect. 5.2. All these authors assume a connection between service value and service design value, where the first becomes a measure for the second. However, no evidence is brought to support this assumption. Focusing on the social value of service design, Lievesley and Yee (2012) then reflected on how to evaluate service

---

[2]www.measuringdesignvalue.eu.

design projects with respect to their Social Return On Investment (SROI). They compare the phases of the service design process with those necessary to calculate the SROI, formulating a hypothesis to embed the SROI evaluation process in service design projects, adapting existing service design tools.

To conclude we cannot forget the importance of measuring design value to justify investments by public and private organizations. This topic applies both to individual organizations and to the service industry as a whole, to evaluate (service) design performances at national and transnational levels. In the attempt to frame the *design economy*, the British Design Council issued a report about the monetary value of design in the UK (Design Council 2015), which justifies the attention toward design competencies and initiatives in the British context. So far, we have not been able to find similar studies specifically focused on service design, even though examples presented in this section can be considered as points of reference for service design also, to structure an evidence-based process that can strengthen its diffusion.

In our opinion opening a debate on service evaluation in service design practice, and starting to operate it pragmatically, can contribute to making more clear, visible and measurable the specific contribution of service design to innovation processes. Our vision is that in a few years, it will then be possible to talk about impacts on people and organizations at the economic and social levels, both at the micro- and macro-scale as it is already happening for the broader design discipline.

6.3   An Interpretive Model of Evaluation in Service Design

Taking inspiration from the models for measuring design value described in the previous section, we propose a framework that describes the possible levels of adoption of evaluation in service design practice. The starting point is the idea of considering evaluation as a learning process capable of determining the value of a service in its complex and systemic nature, considering service both a fully implemented service and a service concept or prototype. As described in Chap. 5, evaluation can represent a crucial element to address service design or redesign interventions. On a different level, as shown in the previous section, it can also assess the contribution (for example in economic terms and in relation to organizational change) of service design as a strategic asset for innovating services.

Reflections made so far provide the basis to build a scheme that describes the possible ways in which evaluation could be adopted in service design practice. In some cases, it can be an element that individually reinforces the research, development and validation steps of the service design process; while in other situations it accompanies the overall design process in an integrated strategy of intervention. In the attempt to interpret such a complex phenomenon and what emerged from the

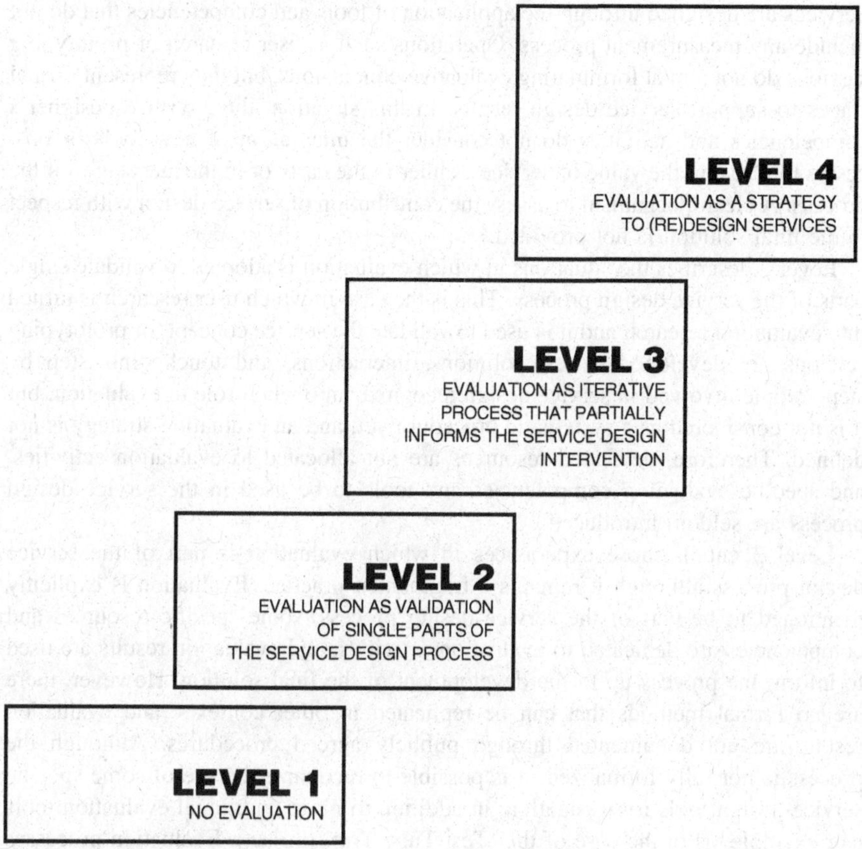

LEVEL 4
EVALUATION AS A STRATEGY
TO (RE)DESIGN SERVICES

LEVEL 3
EVALUATION AS ITERATIVE
PROCESS THAT PARTIALLY
INFORMS THE SERVICE DESIGN
INTERVENTION

LEVEL 2
EVALUATION AS VALIDATION
OF SINGLE PARTS OF
THE SERVICE DESIGN PROCESS

LEVEL 1
NO EVALUATION

Fig. 6.1 Levels of adoption of evaluation in service design practice

exploration on the topic provided in this book, we identify four levels of adoption of evaluation in service design[3] (Fig. 6.1):

1. No evaluation;
2. Evaluation as validation of single parts of the service design process;
3. Evaluation as iterative process that partially informs the service design intervention;
4. Evaluation as a strategy to (re)design services.

Level 1 coincides with the traditional service design approach detailed in Chap. 2, where evaluation is not considered as an explicit activity in the process.

---

[3]This interpretive model must be considered as a first reflection based on the authors' vision, which requires it to be verified through a field inspection addressed at organizations that currently adopt evaluation within their service design practices.

Services are designed through the application of tools and competencies that do not include any measurement process. Operations such as user research or prototyping sessions do not aim at formulating evaluative conclusions, but they represent formal stages to support service design results. In this situation, the , service designer's competencies and activities do not consider the integration of new tools or processes to measure the value of services, either in the early or in the late stages of the process. A clear mechanism to assess the contribution of service design with respect to the final solution is not provided.

Level 2 describes the situations in which evaluation is adopted to validate single parts of the service design process. This is the case in which user research is turned into evaluation research and it is used to validate the service concept, or prototyping sessions are developed to test solutions, interactions, and touchpoints step by step. People involved in service design recognize an explicit role to evaluation, but it is not considered as a systematic procedure yet, and an evaluation strategy is not defined. Therefore, dedicated resources are not allocated to evaluation activities, and specific evaluation competencies and tools to be used in the service design process are seldom introduced.

Level 3 entails those experiences in which evaluation is part of the service design process although it remains a fragmented practice. Evaluation is explicitly mentioned to be part of the service design process, some specific resources and competencies are dedicated to evaluation activities, and evaluation results are used to inform the process up to the development of the final solution. However, there are no formal methods that can be replicated in other contexts, and evaluation results are not documented through publicly agreed procedures. Although the process is not fully formalized, it is possible to recognize the use of some specific service design tools for evaluation, in addition to more traditional evaluation tools (for example, as in the case of the 'Test Tube Trip' project). Evaluation processes are not extemporaneous, but designed through a strategy that defines its scope, tools, schedule, and expected results.

Level 4 describes the situations in which evaluation is fully embedded in service design practice according to an explicit evaluation strategy that guides the entire design process. Evaluation objectives, resources, competencies, tools, and expected results are defined from the early stages of the process, and evaluation is applied in different moments:

- In the beginning, as a preliminary activity useful to identifying the service values that will guide the design process;
- During the design process for monitoring the ongoing results;
- Once the process is concluded, to measure the effects of the service solution on the organization, the communities of users and the context of development, and inform further developments of the solution in the future.

At this level, evaluation is well structured, fully integrated into the service design process, and replicable in other situations. The service designer's skills are extended by incorporating the capability to manage, plan, articulate and design an evaluation journey. New tools are integrated to those traditionally used by service designers or, on the contrary, traditional tools are adapted to the new purpose. In this case, the service designer can also act as an influencer within the community in which it operates, spreading the culture of evaluation within the organization and people involved.

These different levels of adoption of evaluation obviously depend on several factors: the level of maturity of service design in the organization, the various competencies involved, the availability of resources and the organization attitude toward experimentation and innovation. Concerning, in particular, the need to add to new evaluation competencies or to adapt design skills to evaluation some deeper considerations must be made, which are expressed in the next section.

6.4   New Competencies for Service Designers

It is not easy to frame service design as a codified profession. On the one hand, dedicated service design education is becoming increasingly diffused at international level, and thus those who attend a university service design program and obtain a recognized title from it are termed *service designers*. Since the first service design educational program was introduced in 1991 at Köln International School of Design, several other schools have proposed service design as the main subject of Master-level studies or as part of the academic curriculum in interaction design or industrial design (Madano Partnership 2012). On the other hand, service design is still a young discipline, thus meaning that those with diverse backgrounds (or at least a generic design one) can be called service designers, based on their experiences of designing services through a service design approach or using service design tools. Nowadays, in theory, everybody working in the service sector and adopting a creative user-centered approach can be considered a service designer: product and interaction designers, architects, marketing professionals, and ethnographers, among others (Stickdorn and Schneider 2010). As Meroni and Sangiorgi (2011) assert, the primary difficulty for designers—and service designers in particular—is how to communicate the added value they can bring to organizations, especially when they enter a new field of practice or when their contribution overlaps with existing ones, as would be the case with evaluation.

Considering service design as a mindset and a design approach, a multidisciplinary platform of expertise (Moritz 2005) more than a profession codified by a particular academic background, some capabilities characterize service design practitioners, making it possible that they could approach service evaluation as part of the service design process. In tune with Mager's definition (see Sect. 3.4), Moritz (2005) states that service design not only links to the design phase of the service lifecycle but is an ongoing process aimed at addressing organizations' and users' needs, building a bridge between the two within the overall context. He also asserts that service design has the capability of changing the organizational culture, involving people in the project and providing visual explanations that help everybody to understand, share and contribute to the (re)design of the service.

Accordingly, Meroni and Sangiorgi (2011) argue that service designers are capable of designing on four primary levels: service experiences, service policies, service transformations, and service systems. The first level is referred to the capacity of service designers to work within and for organizations to improve existing services or to introduce new ideas. Service policies are more related to a strategic capacity to aligning initiatives to citizens' needs and demands. The third level entails the capacity to facilitate organizations in shifting towards new paradigms (they suggest, in particular, collaborative). Service systems describe how service designers can support organizations in imagining long-term scenarios and future perspectives. Wetter-Edman (2014) also reinforces the idea that service designers can be considered as interpreters of users' experiences, through their ability to use empathic methods as a means of understanding users' contexts.

In general, we can describe some core capabilities for service design and some specific skills that characterize the service designer's mindset and profession. The ability to listen to and empathically understand people (Wetter-Edman 2014) is one of the features that service designers apply to engage people and gather information from their behaviors. In some circumstances this also means playing a mediator role, facilitating participatory and collaborative activities (Tan 2012; Napier 2015), and stimulating knowledge sharing among different stakeholders. Service designers are often involved in guiding and coaching creative initiatives (such as workshops, jams, hackathons, sprints). The ability to collaborate with different stakeholders and the capacity to enable them playing an active role is fundamental to envision solutions, to establish a sense of community, and to share values. Moreover, service designers have to deal with data, suggestions, and insights from research making them useful, clear, and shareable among designers and non-designers. This process is not only related to analytical skills, but also to the capacity to think strategically and to transform data into solutions. Communication and visualization skills support these initiatives, making complex stories, processes, and relationships visible and accessible to a large number of stakeholders. Service designers can create proper storytelling capable of inspiring and guiding research and creative processes on the one hand, and making these processes and their results concrete and measurable on the other hand. The capacity to make ideas tangible is intrinsically part of any designer capacity, using specific tools to visualize ideas, concepts and prototypes and to deliver and implement the experiences designed on paper (Morelli and Tollestrup 2006).

Services are intrinsically complex, so designing services also requires a system thinking capability (Darzentas and Darzentas 2014). Service designers are always faced with systems of actors, processes, interactions, emotions, physical and digital elements. All these components need to be coherent and harmonious and respondent to the user and the provider expectations, which means considering services as systems and not as the sum of single elements. To transform service concepts into real services different design abilities are required. As a matter of fact, also product designers, interaction designers and interior designers can intervene in service design projects (Stickdorn and Schneider 2010). No less important is the ability to understand and manage business dynamics. The 'Business Model Canvas' and the

'Value Proposition Canvas' have become essential tools for service designers to assess the appropriateness of service ideas with respect to target users and the market, to better define the development strategies, and to dialogue with business experts using a common language.

In his recent book, Michlewski (2015) describes five aspects characterizing the design attitude from a management perspective. Among these is the capacity to face ambiguity, namely the ability to feel comfortable with uncertainty and discontinuity. This is related to the idea of embracing risks and adopting abductive processes. Empathizing with users is another important attitude, described as the ability to see an object, an experience or a service through another person's point of view. For sure, empathy, the capacity to listen to users and to wear users' shoes, is one of the most important characteristics recognized for service designers (Kouprie and Sleeswijk Visser 2009; New and Kimbell 2013; Reason et al. 2016). The third attitude is related to the aesthetic dimension (in its sophisticated way to include beauty, emotions, consistency of the solutions, and so on) and the designer capacity to use all the five senses to develop products and services that create powerful connections, brand accountability, and multilevel experiences for the final users. Another characteristic that defines the designer attitude is the capacity to play, namely to experiment and embrace playfulness to bring things to life. This is connected to the ideation phase, to the prototyping one, and to the capacity to visualize data through processes and artifacts that make information easier to understand, beautiful and impactful. Another well-explored attitude consists of facing complexity and new meanings (Boland and Collopy 2009; Weick et al. 2005) to support innovation by mashing-up cultural, technical, and aesthetic elements of products and services.

Referring to evaluation, it is then important to understand what are the skills of the evaluator, even if, in this case also, it is not possible to recognize a codified profession yet. In the US, in particular, the first evaluators were social scientists (sociologists, economists, educators, psychologists) who tried to apply their expertise to the necessities of the real world (Shaw et al. 2006). The increasing demand for evaluation (especially in the public sector) gave birth to specialized consultancies, as well as the rise of practitioners describing themselves as evaluators. Despite the increase in dedicated university programs, evaluation is struggling to be recognized as a profession. There is significant heterogeneity in the disciplinary and professional training, and evaluators are first of all practitioners with a long experience in doing the evaluation. Although knowledge of the theory and methodology of program evaluation is essential for conducting a proper evaluation, a significant amount of knowledge concerns the target problem area (such as health, crime, education, etc.) and the specific elements of the evaluation context (Rossi et al. 2004). According to Rossi et al. (2004), two extremes can be recognized: at the most complex level, evaluation can be so technical and sophisticated in conception that it requires the participation of highly trained specialists with knowledge of theory, data collection methods, and statistical techniques. At the other extreme, evaluation can be so easy that persons of modest expertise and experience can carry it out. Nevertheless, even considering this broad range of situations, it is possible to infer some characteristics the evaluator should have (Bezzi 2007):

- *Research approach*: the evaluator should know research methodologies, tech-niques, and how to deal with epistemological issues. This aspect is crucial to avoid running into useless technicalities or problems created by not selecting the appropriate techniques from the start;
- *Communication skills*: the evaluator should have an empathic attitude, to interact with decision-makers and other actors involved, to be able to interpret and address his/her client requests toward a clear and shared evaluation man-date, and to influence the use of evaluation results authoritatively;
- *Critical attitude*: the evaluator should be curious and open to critique, also concerning the results obtained. He or she should distance himself/herself from the common way of thinking to interpret data and produce shared evidence objectively.

To sum up, the evaluator does not necessarily possess technical skills, because his/her role is first of all to design the strategy and interpret results. For example, an evaluator in charge of assessing a new urban plan will probably require the involvement of a town planner to use the proper techniques concerning urban planning evaluation.

Making a comparison between the evaluators' skills and those of the service designer, we can identify some overlaps between the competencies of the two figures:

- Observation of current situations from the user and the organization point of view;
- Critique and interpretation of insights emerging from observation;
- Transformation of problems into solutions that generate value for both the organization and the user;
- Visualization of solutions to make them more understandable for the actors involved.

Considering the service evaluation definition proposed in this book, and given the need to design an evaluation strategy, we can affirm that service design practitioners are endowed with the necessary characteristics to address this challenge, at least at the strategic level. Thanks to their attitudes to system and visual thinking, which casts them as interpreters, transformers, and, in a way, educators, they can be considered *evaluation designers*. They can contribute to identifying evaluation objects and objectives, to plan and conduct data collection, and to interpret results, eventually making data visualization more accessible. Imagining an evolution of the figure of the service designer professional, it is, therefore, possible to imagine an integration between the two profiles. Nevertheless, what is proposed is not to make service designers substitute expert evaluators, but rather to make service designers approach evaluation as an essential activity for service design, completing and supporting the specialist knowledge of expert evaluators. This outlines an articu-lated profile capable of working in strategic areas of research and innovation through the adoption of holistic and human-centered processes able to produce high-quality deliverables. Today, organizations and design agencies that deal with service design do not have specific offers dedicated to the evaluation. Referring to the service design practice, it is thus necessary to further experiment with this approach, integrating the service designer's skills with the ability to design, manage and carry out an evaluation process. This involves an extension of the service design practice to other areas of expertise, requiring a reasonable modification of existing approaches and tools. Accordingly, it becomes fundamental to strengthen the conceptual reflection on these issues within the scientific design community and with the extra-disciplinary contexts that address, at various levels, the evaluation topic.

6.5   Service Design and Service Evaluation: Open Issues

Being most of the subjects treated in this book are a novelty in the field, and the interpretations proposed based on conceptual hypotheses, we cannot avoid concluding our argumentation with some open issues that should be investigated by future research and experimentations.

We are facing *the great transformation*,[4] where the rapid change of product-service design, as well as production and distribution processes, have significantly enlarged the demand for a service design approach. In the last decade, publications on service design have multiplied, the educational offer has grown as well as research activities exploring various topics and new scenarios for the discipline. This is the consequence of the fact that private and public organizations need to continuously improve the quality of experiences offered, to control and optimize the use of resources, and to generate innovative solutions that answer to emerging and evolving needs. However, on the other hand, we are facing the globalization process, which causes a generalized proliferation and commodification of service solutions generated by a simplified service design approach, which materializes low-quality outputs and scarce diversity.

Service design is ready for new challenges, which necessarily require critically inspecting and rethinking what has finally been recognized as an established approach supported by consolidated tools. In this situation, it is useful, from our point of view, to strengthen the academic research and training by exploring new territories and new interdisciplinary connections, to contribute to the development and evolution of the discipline and the related professional practice. To provide through evidence a better understanding of service complexity, we proposed evaluation as a powerful instrument in the hands of service design practitioners, to be used in tandem with traditional design research methods. The idea is to spread a culture of service evaluation integrated within the design practice that goes beyond

---

[4]*The Great Transformation* is a book by the economist Karl Polanyi published in 1944 describing the radical social and political changes that characterized England's contexts shifting from a premodern economy to a market economy.

the administration of surveys and interviews or accountability purposes, becoming a processual activity coherently designed within the overall service design process. Nonetheless, this must be considered as the first step of a long journey, which should lead to the building of shared theories and practices useful to systematize and codify what is presented here as a first proposal. The next step will consist in validating or confuting our hypotheses within the academic community and among practitioners, through the application of the integrated service design and evaluation process described in Sect. 5.3, and guidelines presented in Sect. 5.5. Moreover, some of the issues we have not been able to explore in this book should be soon taken into consideration. Among others that the reader could certainly identify we highlight the following:

1. **The difference between the design and evaluation of public and private services**. Designing better services, as suggested by the title of this book, is a purpose for both public and private organizations. Given the necessity to frame the topic of service evaluation in service design practice prior to exploring its peculiarities when applied to specific service sectors, we decided not to make a distinction between the evaluation of public and private services. Nevertheless, we cannot ignore that public and private sectors explicitly address different value systems. Public services aim to satisfy fundamental welfare needs, whereas private services aim to attain competitive advantage in the market. This difference inevitably affects how services are designed and evaluated. As we can infer from program evaluation literature, public services are usually evaluated from a wider policy perspective and interventions emerging from evaluation affect a broader system of public actions. In the public sector, there is greater awareness of the importance of evaluation as a learning and transformation process, demonstrated by the fact that specific resources and competencies are increasingly embedded within public bodies and institutions. The same happens from a design perspective, as shown by the emergence of the new stream of study and practice called *design for policy*, which establishes a connection between policies, programs and services and the need for them to be coherently designed (Bason 2010, 2014). For these reasons, for what concerns public services, we believe the link between program evaluation, service design, and *design for policy* needs to be explored, providing a better framing of the concept of service evaluation in the public context, and adapting the model proposed to the policy design process too.

2. **Extending service design expertise to evaluation**. The legitimacy for service design to enter into evaluation processes that are usually appointed to special-ized professional figures is a delicate issue, both at the academic and profes-sional levels. Given the similarities between some of the competencies of the service designer and that of the expert evaluator, what we suggest is to jointly approach the integration of their expertise. Evaluation experts should accom-pany service designers first to develop an evaluation culture, and then to pragmatically set up and conduct an evaluation strategy, actively participating in service design projects. This would allow better understanding the actual

contribution evaluation could bring to service design, both concerning service evaluation and service design evaluation, validating or refining the hypothesis presented in this book. For this to happen a deeper investigation needs to be undertaken on the roles these figures could play and what their collaboration would imply in each step of the process. Making this process systematic, operative and replicable so that organizations can easily adopt it is a challenge strictly connected to competencies that need to be involved. Moreover, for the acknowledgment and diffusion of an evaluation component in service design practice a reflection on the education of new professional profiles is required. This consists of imagining renovated training programs for service design, which also include theoretical learning and experimentations on the topic. As stated above, the purpose is not to substitute the role of expert evaluators, but rather to acquire a mindset, methods and tools to enable a fruitful collaboration and reinforce the spreading of the discipline.

3. **Building a system of metrics to measure service and service design value**. In Chap. 5 we defined the elements to be considered when designing a service evaluation strategy (the evaluation objectives, the object of evaluation, the perspective of evaluation, the time of evaluation) , specifying that for what concerns evaluation objectives and objects it is not possible to compile a list of predefined options among which to choose. We described some possible categorizations emerging from literature, which represent an attempt to simplify service complexity. Nevertheless, specific objects and objectives can be found only in relation to specific contexts of evaluation. For this reason, it is necessary to investigate how to identify them, and to elaborate related indicators and metrics necessary for their measurement, both from the user and the organization perspective. It is argued that this issue could be resolved through experience, that is, the application of the model to a large number of cases. This would probably allow the identification, at least, of clusters of evaluation objects characterizing specific service sectors (for example transportation services, educational services, hospitality services, etc.). Based on the same logic, the investigation should then concentrate on discovering value objectives, objects, and metrics for the evaluation of service design. What has been done so far for measuring the value of design is for sure a good starting point, but some work still needs to done to demonstrate with objectivity the contribution (and the impact) of service design to innovation, or, more specifically, in economic, social, environmental and political terms. This remains a great challenge and opportunity for research and experimentation.

Many other issues could be raised, which we have not discussed so far, but that certainly deserve attention, such as the role of participatory approaches in the process of evaluation, or the peculiarities of the evaluation of the digital components of services, just to mention a few. There is room for significant improvement from several points of view, which requires shifting the boundaries of service design. If future problems and situations cannot be predicted with certainty, they can be understood by looking at what is already here and change them for the better. Our wish is that the debate on these topics within the discipline could grow and that the relationship with other disciplines could be strengthened. We need to be visionary today for things to be ordinary in the next future.

References

Andreassen TW, Kristensson P, Lervik-Olsen L, Parasuraman A, McColl-Kennedy JR, Edvardsson B, Colurcio M (2016) Linking service design to value creation and service research. J Serv Manage 27(1):21–29

Bason C (2010) Leading public sector innovation: co-creating for a better society. Policy Press, Bristol

Bason C (2014) Design for policy. Gower Publishing, Aldershot

Bezzi C (2007) Cos'è la valutazione. Un'introduzione ai concetti, le parole chiave e i problemi metodologici. Franco Angeli, Milano

Bezzi C (2010) Il nuovo disegno della ricerca valutativa. Franco Angeli, Milano

Boland R, Collopy F (eds) (2009) Managing as designing. Stanford University Press, Stanford

Borja de Mozota B (2002) Design and competitive edge: a model for design management excellence in European SMEs. Des Manage J 2(1):88–103

Borja de Mozota B (2011) Design strategic value revisited: a dynamic theory for design as organizational function. In: Cooper R, Junginger S, Lockwood T (eds) The handbook of design management. Berg Publishers, Oxford, pp 276–293

Buchanan C, Junginger S, Terrey N (2017) Service design in policy making. In: Sangiorgi D, Prendiville A (eds) Designing for service: key issues and new directions. Bloomsbury, London, pp 183–198

Darzentas J, Darzentas J (2014) Systems thinking in design: service design and self-services. FORMakademisk 7(4):1–18

Design Council (2015) The design economy; the value of design to the UK. Available via Design Council. http://www.designcouncil.org.uk/resources/report/design-economy-report. Accessed 3 June 2017

European Commission (2009) Design as a driver of user-centred innovation. http://ec.europa.eu/DocsRoom/documents/2583/attachments/1/translations/en/renditions/native. Accessed 30 May 2017

European Commission (2013) Implementing an action plan for design-driven innovation. http://ec.europa.eu/DocsRoom/documents/13203/attachments/1/translations/en/renditions/native. Accessed 30 May 2017

Holmlid S, Wetter-Edman K, Edvardsson B (2017) Breaking free from NSD: design and service beyond new service development. In: Sangiorgi D, Prendiville A (eds) Designing for service: key issues and new directions. Bloomsbury, London, pp 95–104

Junginger S (2009) Design in the organization: parts and wholes. Des Res J 2(9):23–29

Junginger S (2013) Design and innovation in the public sector: matters of design in policy making and policy implementation. In: Proceedings of the 10th European academy of design conference, Gothenburg University, Gothenburg, 17–19 Apr 2013

Kimbell L (2015) Applying design approaches to policy making: discovering Policy Lab. https://researchingdesignforpolicy.files.wordpress.com/2015/10/kimbell_policylab_report.pdf. Accessed 21 May 2017

Kimbell L (2016) Design in the time of policy problems. In: Proceedings of DRS 2016, Brighton, 27–30 June 2016

Kimbell L, Blomberg J (2017) The object of service design. In: Sangiorgi D, Prendiville A (eds) Designing for service: key issues and new directions. Bloomsbury, London, pp 106–120

Kootstra GL (2009) The incorporation of design management in today's business practices. http://www.bcd.es/site/unitFiles/2585/DME_Survey09-darrera%20versi%C3%B3.pdf. Accessed 1 June 2017

Kouprie M, Sleeswijk Visser F (2009) A framework for empathy in design: stepping into and out of the user's life. J Eng Des 20(5):437–448

Lievesley M, Yee J (2012) Valuing service design: lessons from SROI. In: Proceedings of DRS 2012, Bangkok, 1–4 Jul 2012

Løvlie L, Downs C, Reason B (2008) Bottom-line experiences: measuring the value of design in service. Des Manag Rev 19(1):73–79

Madano Partnership (2012) Scoping study on service design. Available via Design Council. https://www.designcouncil.org.uk/sites/default/files/asset/document/Scoping%20Study%20on%20Service%20Design%20Final_website%20summary_v2.pdf. Accessed 18 Sept 2016

Manschot M, Sleeswijk Visser F (2011) Experience-value: a framework for determining values in service design approaches. In: Proceedings of IASDR 2011, Delft, 31 Oct–4 Nov 2011

Manzini E (2014) Design and policies for collaborative services. In: Bason C (ed) Design for policy. Gower Publishing, Farnham, pp 103–112

McNabola A, Moseley J, Reed B, Bisgaard T, Jossiasen AD, Melander C, Whicher A, Hytönen J, Schultz O (2013) Design for public good. Available via Design Council. http://www.designcouncil.org.uk/sites/default/files/asset/document/Design%20for%20Public%20Good.pdf. Accessed 15 May 2017

Meroni A, Sangiorgi D (2011) Design for services. Gower Publishing, Surrey

Micheli P (2014) Leading business by design: why and how business leaders invest in design. Available via Design Council. https://www.designcouncil.org.uk/sites/default/files/asset/document/dc_lbbd_report_08.11.13_FA_LORES.pdf. Accessed 3 June 2017

Michlewski K (2015) Design attitude. Gower Publishing, Farnham

Morelli N, Tollestrup C (2006) New representation techniques for designing in a systematic perspective. In: Proceedings of the 8th international conference on engineering and product design education, Salzburg, 7–8 Sept 2006

Moritz S (2005) Service design. Practical access to an evolving field. Available via http://stefan-moritz.com/welcome/Service_Design_files/Practical%20Access%20to%20Service%20Design.pdf. Accessed 2 Feb 2016

Mortati M, Villari B (2016) Design capabilities and business innovation. In: DeFilippi R, Rieple A, Wikström P (eds) International perspectives on business innovation. Edward Elgar Publishing, Northampton, pp 256–275

Mortati M, Villari B, Maffei S (2014) Design capabilities for value creation. In: Proceedings of the 19th DMI: Academic design management conference, London, 2–4 Sept 2014

Moultrie J, Livesey F (2009) International design scoreboard: initial indicators of international design capabilities. http://www.idi-design.ie/content/files/InternationalDesignScoreboard.pdf. Accessed 3 Oct 2015

Napier P (2015) Design facilitation: training the designer of today. In: Proceedings of cumulus conference, Milan, 3–7 June 2015

New S, Kimbell L (2013) Chimps, designers, consultants and empathy: a "Theory of Mind" for service design. In: Proceedings of the 2nd Cambridge academic design management conference, Cambridge, 4–5 Sept 2013

Palumbo M (2001) Il processo di valutazione. Decidere, programmare, valutare. Franco Angeli, Milano

Ramlau U, Melander C (2004) In Denmark, design tops the agenda. Des Manage Rev 15(4):48–54

Reason B, Løvlie L, Brand FLuM (2016) Service design for business: a practical guide to optimizing the customer experience. Wiley, New York

Rossi PH, Freeman HE, Lipsey MW (2004) Evaluation. A systematic approach, 7th edn. Sage, Thousand Oaks

Shaw IF, Greene JC, Melvin MM (2006) The Sage handbook of evaluation. Sage, London

Stickdorn M, Schneider J (2010) This is service design thinking. BIS Publishers, Amsterdam

Tan L (2012) Understanding the different roles of the designer in design for social good. A study of design methodology in the DOTT 07 (Designs of the Time 2007) Projects. Dissertation, Northumbria University

Thomson M, Koskinen T (2012) Design for growth and prosperity. http://european-designinnovation.eu/wp-content/uploads/2012/09/Design_for_Growth_and_Prosperity_.pdf. Accessed 17 Mar 2017

Villari B (2012) Design per il territorio. Un approccio community centred. FrancoAngeli, Milano

Weick KE, Sutcliffe KM, Obstfeld D (2005) Organizing and the process of sensemaking. Organ Sci 16(4):409–421

Westcott et al (2013) The DMI design value scorecard: a new design measurement and management model. Des Manage Rev 23(4):10–16

Wetter-Edman K (2014) Design for service. A framework for articulating designers' contribution as interpreter of users' experience. University of Gothenburg, Gothenburb

Index

© The Author(s) 2018
F. Foglieni et al., *Designing Better Services*, PoliMI SpringerBriefs,
https://doi.org/10.1007/978-3-319-63179-0

Printed in the United States
By Bookmasters